NETWORK SCIENCE

Committee on Network Science for Future Army Applications

Board on Army Science and Technology
Division on Engineering and Physical Sciences

NATIONAL RESEARCH COUNCIL
OF THE NATIONAL ACADEMIES

THE NATIONAL ACADEMIES PRESS
Washington, D.C.
www.nap.edu

THE NATIONAL ACADEMIES PRESS   500 Fifth Street, N.W.   Washington, DC 20001

NOTICE: The project that is the subject of this report was approved by the Governing Board of the National Research Council, whose members are drawn from the councils of the National Academy of Sciences, the National Academy of Engineering, and the Institute of Medicine. The members of the committee responsible for the report were chosen for their special competences and with regard for appropriate balance.

This study was supported by Contract/Grant No. DAAD19-03-D-0002, between the National Academy of Sciences and the Department of the Army. Any opinions, findings, conclusions, or recommendations expressed in this publication are those of the author(s) and do not necessarily reflect the views of the organizations that provided support for the project.

International Standard Book Number 0-309-10026-7 (Book)
International Standard Book Number 0-309-65388-6 (PDF)
Library of Congress Control Number 2005936575

Additional copies of this report are available from the National Academies Press, 500 Fifth Street, N.W., Lockbox 285, Washington, DC 20055; (800) 624-6242 or (202) 334-3313 (in the Washington metropolitan area); Internet, http://www.nap.edu.

Copyright 2005 by the National Academy of Sciences. All rights reserved.

Printed in the United States of America

THE NATIONAL ACADEMIES
Advisers to the Nation on Science, Engineering, and Medicine

The **National Academy of Sciences** is a private, nonprofit, self-perpetuating society of distinguished scholars engaged in scientific and engineering research, dedicated to the furtherance of science and technology and to their use for the general welfare. Upon the authority of the charter granted to it by the Congress in 1863, the Academy has a mandate that requires it to advise the federal government on scientific and technical matters. Dr. Ralph J. Cicerone is president of the National Academy of Sciences.

The **National Academy of Engineering** was established in 1964, under the charter of the National Academy of Sciences, as a parallel organization of outstanding engineers. It is autonomous in its administration and in the selection of its members, sharing with the National Academy of Sciences the responsibility for advising the federal government. The National Academy of Engineering also sponsors engineering programs aimed at meeting national needs, encourages education and research, and recognizes the superior achievements of engineers. Dr. Wm. A. Wulf is president of the National Academy of Engineering.

The **Institute of Medicine** was established in 1970 by the National Academy of Sciences to secure the services of eminent members of appropriate professions in the examination of policy matters pertaining to the health of the public. The Institute acts under the responsibility given to the National Academy of Sciences by its congressional charter to be an adviser to the federal government and, upon its own initiative, to identify issues of medical care, research, and education. Dr. Harvey V. Fineberg is president of the Institute of Medicine.

The **National Research Council** was organized by the National Academy of Sciences in 1916 to associate the broad community of science and technology with the Academy's purposes of furthering knowledge and advising the federal government. Functioning in accordance with general policies determined by the Academy, the Council has become the principal operating agency of both the National Academy of Sciences and the National Academy of Engineering in providing services to the government, the public, and the scientific and engineering communities. The Council is administered jointly by both Academies and the Institute of Medicine. Dr. Ralph J. Cicerone and Dr. Wm. A. Wulf are chair and vice chair, respectively, of the National Research Council.

www.national-academies.org

COMMITTEE ON NETWORK SCIENCE FOR FUTURE ARMY APPLICATIONS

CHARLES B. DUKE, *Chair*, Xerox Innovation Group, Webster, New York
JOHN E. HOPCROFT, *Vice Chair*, Cornell University, Ithaca, New York
ADAM P. ARKIN, Lawrence Berkeley National Laboratory, Berkeley, California
ROBERT E. ARMSTRONG, National Defense University, Washington, D.C.
ALBERT-LASZLO BARABASI, University of Notre Dame, Notre Dame, Indiana
RONALD J. BRACHMAN, Defense Advanced Research Projects Agency, Arlington, Virginia
NORVAL L. BROOME, MITRE Corporation (retired), Suffolk, Virginia
STAN DAVIS, Brookline, Massachusetts
RICHARD A. De MILLO, Georgia Institute of Technology, Atlanta
WILLIAM J. HILSMAN, Institute for Defense Analyses, Philadelphia, Pennsylvania
WILL E. LELAND, Telcordia Technologies, Inc., Piscataway, New Jersey
THOMAS W. MALONE, Massachusetts Institute of Technology, Cambridge
RICHARD M. MURRAY, California Institute of Technology, Pasadena
JACK PELLICCI, Oracle Public Sector, Reston, Virginia
PAMELA A. SILVER, Harvard Medical School, Boston, Massachusetts
PAUL K. VAN RIPER, LTG, United States Marine Corps (retired), Williamsburg, Virginia
DUNCAN J. WATTS, Columbia University, New York

Staff

ROBERT J. LOVE, Study Director
NIA D. JOHNSON, Research Associate
TOMEKA N. GILBERT, Senior Program Assistant (until May 23, 2005)
LEON A. JAMES, Senior Program Assistant (after May 23, 2005)

BOARD ON ARMY SCIENCE AND TECHNOLOGY

JOHN E. MILLER, *Chair*, L3 Communications Corporation, Reston, Virginia
HENRY J. HATCH, *Vice Chair*, Army Chief of Engineers (retired), Oakton, Virginia
SETH BONDER, The Bonder Group, Ann Arbor, Michigan
JOSEPH V. BRADDOCK, The Potomac Foundation, McLean, Virginia
NORVAL L. BROOME, MITRE Corporation (retired), Suffolk, Virginia
ROBERT L. CATTOI, Rockwell International (retired), Dallas, Texas
DARRELL W. COLLIER, U.S. Army Space and Missile Defense Command (retired), Leander, Texas
ALAN H. EPSTEIN, Massachusetts Institute of Technology, Cambridge
ROBERT R. EVERETT, MITRE Corporation (retired), New Seabury, Massachusetts
PATRICK F. FLYNN, Cummins Engine Company, Inc. (retired), Columbus, Indiana
WILLIAM R. GRAHAM, National Security Research, Inc., Arlington, Virginia
PETER F. GREEN, University of Michigan, Ann Arbor
EDWARD J. HAUG, University of Iowa, Iowa City
M. FREDERICK HAWTHORNE, University of California, Los Angeles
CLARENCE W. KITCHENS, Science Applications International Corporation, Vienna, Virginia
ROGER A. KRONE, Boeing Integrated Defense Systems, Philadelphia, Pennsylvania
JOHN W. LYONS, U.S. Army Research Laboratory (retired), Ellicott City, Maryland
MALCOLM R. O'NEILL, Lockheed Martin Corporation, Bethesda, Maryland
EDWARD K. REEDY, Georgia Tech Research Institute (retired), Atlanta
DENNIS J. REIMER, DFI International, Washington, D.C.
WALTER D. SINCOSKIE, Telcordia Technologies, Inc., Piscataway, New Jersey
JUDITH L. SWAIN, University of California, San Diego
WILLIAM R. SWARTOUT, Institute for Creative Technologies, Marina del Rey, California
EDWIN L. THOMAS, Massachusetts Institute of Technology, Cambridge
BARRY M. TROST, Stanford University, Stanford, California

Staff

BRUCE A. BRAUN, Director
WILLIAM E. CAMPBELL, Manager, Program Operations
CHRIS JONES, Financial Associate
ROBERT J. LOVE, Senior Program Officer
MARGARET NOVACK, Senior Program Officer
HARRISON T. PANNELLA, Senior Program Officer
DONALD L. SIEBENALER, Senior Program Officer
DEANNA P. SPARGER, Program Administrative Coordinator

Preface

This study was an exercise in coping with complexity. The subject matter is complex. Important networks like the Internet and the power grid are becoming ever larger, encompassing up to hundreds of millions if not billions of nodes. They exhibit complex and often dynamic patterns of links between the nodes. Networks interact with one another and are recursive. Social networks are built upon information networks which are built upon communications networks which in turn are built on physical networks. Moreover, this layered structure of interacting networks built on top of other networks is reflected directly in the diversity of communities that study networks: sociologists, management theorists, warfare strategists, economists, biologists, chemists, physicists, and a wide variety of engineers. Getting such a diverse group to agree on a common core of knowledge about networks, i.e., the content of network science, is a significant challenge. Last but by no means least, the customer community for this study is equally diverse. Military planners and strategists, operational commanders ("warfighters"), logistics commanders, and R&D managers each have their own points of view on what network science ought to provide in order to be useful to the military.

The extent to which the committee did manage to cope successfully with these complexities will be judged by you, the reader. In order to comprehend the topical scope, committee members were selected who are actively publishing experts on physical, biological, engineered, and social networks. Systematic efforts at outreach to interested communities were undertaken, including a survey of extant courses on networks and a questionnaire sent to members of as diverse a group of communities as the committee could identify. Committee members also were selected to encompass various constituencies in the military that have an interest in the design, procurement, deployment, and use of networks. Representatives of each of these groups made presentations to the committee. Value creation scenarios were prepared to address the concerns of these constituencies. Thus, the composition of the committee, the data that it collected, and the analyses that it generated are broadly representative of the inherent complexities of the subject of the study.

The committee was able to lay out the scope of the topic, organize an overview of the diverse streams of activity and knowledge into a synthetic whole, and survey the sorts of options that the Army might want to explore to create value from investments in network science. As a result, it is my hope that this report will broaden the horizons of its readers, stimulate them to think about the role of network science in today's connected world, and, hopefully, act upon their enhanced understanding of this role.

The committee learned three major things of overarching importance about the role of networks in modern society and the availability of the knowledge necessary to create and operate them. First, networks lie at the core of the economic, political, and social fabric of the 21st century. The demand for structured knowledge that can be used to design, procure, and operate networks is ubiquitous and growing rapidly. Moreover, social and communications networks lie at the core of both conventional military operations and the war on terrorism. Thus, investment in network science is both a strategic and urgent national priority.

Second, the current state of knowledge about the structure, dynamics, and behaviors of both large infrastructure networks and vital social networks at all scales is primitive. A lot is known about the design, construction, and use of the components of physical networks. The science of integrating these components into large, complex, interacting networks that are robust and whose behaviors are predictable is uncharted ground. Communications networks that are being built today exhibit unpredictable behavior and robustness. For social networks, even the characteristics of the components are largely unexplored. The development of predictive models of the behavior of large complex networks is difficult. It is basically an unsolved problem that will require focused attention from the best brains in the nation to make significant progress on it.

Third, the United States is not on track to consolidate the information that already exists about the science of large, complex networks, much less to develop the knowledge that will be needed to design the networks envisaged by the military to realize futuristic warfare concepts like network-centric operations. Current research on networks is highly fragmented, usually conducted in disciplinary settings. The committee did observe an encouraging preliminary consensus on the part of practitioners about the broad outlines of the core of knowledge that allows them to practice their art in a wide variety of applications areas. Nevertheless, individual researchers are naturally more interested in marketing their own work than in collaborating on larger projects of the scope that would have a realistic chance to impact the Army's aspirations. Major changes in the funding and organization of activities on network science are required before the knowledge that can realistically be expected from research in this area will be available in a form that is useful for the design and procurement of the capabilities envisaged by the Army.

The committee does not expect its report to change any of these things. It does, however, aspire to articulate its learnings clearly and to document the data and analysis on which they are based. It also aspires to provide specific answers to the questions in the statement of task. I hope that the readers will find that these aspirations were accomplished.

Charles B. Duke, *Chair*
Committee on Network Science
for Future Army Applications

Acknowledgments

The committee thanks the organizations and guest speakers that provided support. The presentations from the military on network-centric operations (NCO), on the procurement and deployment of operational capabilities of NCO in the field, and on the Army's R&D portfolio of network-related research were especially helpful.

The committee also thanks the various academic and military researchers with whom it conducted personal interviews over the telephone. Their candid comments were instrumental in the committee's achieving a realistic understanding of the complexities of current research on networks.

The committee is deeply grateful to Katy Börner of Indiana University for her analysis of the data acquired from our outreach questionnaire and her permission to use this material in its report.

The excellent support of the National Research Council staff is especially appreciated. Special thanks go to Bob Love, who worked closely with the chair and vice chair during the entire study process. The cheerful and effective assistance of Tomeka Gilbert, Nia Johnson, Deborah Kuzmanovic, and Leon James was indispensable to accomplishing this study.

Acknowledgment of Reviewers

This report has been reviewed in draft form by individuals chosen for their diverse perspectives and technical expertise, in accordance with procedures approved by the NRC Report Review Committee. The purpose of this independent review is to provide candid and critical comments that will assist the institution in making its published report as sound as possible and to ensure that the report meets institutional standards for objectivity, evidence, and responsiveness to the study charge. The review comments and draft manuscript remain confidential to protect the integrity of the deliberative process. We wish to thank the following individuals for their review of this report:

Anthony Ephremides, University of Maryland,
Gerald J. Iafrate, North Carolina State University,
Leonard Kleinrock, NAE, University of California, Los Angeles,
Scott E. Page, University of Michigan,
Lawrence G. Roberts, NAE, Anagran, Inc.,
Alan B. Salisbury, U.S. Army (retired), and
Judith L. Swain, IOM, University of California, San Diego.

Although the reviewers listed above have provided many constructive comments and suggestions, they were not asked to endorse the conclusions or recommendations nor did they see the final draft of the report before its release. The review of this report was overseen by Stewart D. Personick, NAE, Drexel University. Appointed by the National Research Council, he was responsible for making certain that an independent examination of this report was carried out in accordance with institutional procedures and that all review comments were carefully considered. Responsibility for the final content of this report rests entirely with the authoring committee and the institution.

Contents

EXECUTIVE SUMMARY 1

1 INTRODUCTION 7
Scope of the Study, 8
Study Approach and Constraints, 8
Report Organization, 9
References, 10

2 NETWORKS AND NETWORK RESEARCH IN THE 21ST CENTURY 11
References, 18

3 NETWORKS AND THE MILITARY 19
Networks and the Army, 19
Network-Centric Warfare and Network-Centric Operations, 19
Challenges, 22
Optimizing Warfighting Organizations, 22
Network Research of Special Interest to the Military, 24
References, 25

4 THE DEFINITION AND PROMISE OF NETWORK SCIENCE 26
What Is Network Science?, 26
Positioning of Network Science, 28
References, 29

5 THE CONTENT OF NETWORK SCIENCE 30
How Do We Know?, 30
Content, 30
References, 32

6 STATUS AND CHALLENGES OF NETWORK SCIENCE 33
Key Messages, 33
Questionnaire Process, 33
The Respondents, 33
Dissenting Voices, 34
Defining the Field, 34
    Attributes of a Network, 34
    Derived Properties of Networks, 35
    Future Evolution of the Definition of Network Science, 36

Research Challenges, 36
The Social Structure of Network Science, 37
Reference, 38

7　CREATING VALUE FROM NETWORK SCIENCE:　　39
　　SCOPE OF THE OPPORTUNITY
　　Creating Economic Value from Research Knowledge, 39
　　Scenarios for Value Creation, 39
　　　　Scenario 1, Building the Base, 39
　　　　Scenario 2, Next-Generation R&D, 40
　　　　Scenario 3, Creating a Robust Network-centric Warfare/Operations Capability, 41
　　　　Implication of the Scenarios, 42
　　Findings from Scenario 1, 42
　　Findings from Scenarios 2 and 3, 44
　　References, 45

8　CONCLUSIONS AND RECOMMENDATIONS　　46
　　Overarching Conclusions, 46
　　Specific Conclusions, 48
　　Recommendations, 49

APPENDIXES

A	BIOGRAPHICAL SKETCHES OF COMMITTEE MEMBERS	55
B	COMMITTEE MEETINGS AND OTHER ACTIVITIES	58
C	CONTENT OF NETWORK SCIENCE COURSES	60
D	QUESTIONNAIRE DATA	65
E	OPPORTUNITIES FOR CREATING VALUE FROM NETWORK SCIENCE	93
F	RECOMMENDED READING LIST	107

Figures, Tables, and Boxes

FIGURES

2-1   Number of papers with the term "complex network" in the title, 15
2-2   Magazines and journals with articles on networks, 16

3-2-1 Representative activities and networks involved in responses to a bioterrorist attack, 23

6-1   Reasons for saying there is no field of network science, 35
6-2   Share of responses that mention an attribute, 35
6-3   Responses identifying driving applications, 36
6-4   Major research challenges, 36
6-5   Relationships among invitees, respondents, and collaborators, 37
6-6   Network science researchers network, 37

D-1   New names by response ID, 72
D-2   Countries where respondents were located, 73
D-3   States where respondents were located, 75
D-4   Fields selected by respondents, 77
D-5   Most frequently mentioned fields, 78
D-6   Reasons for saying there is no field of network science, 79
D-7   Responses identifying network attributes, 80
D-8   Derived properties of networks mentioned by respondents, 81
D-9   Driving applications identified by respondents, 84
D-10 Number of responses to driving applications question, 84
D-11 Major research challenges, 86
D-2-1 Relationships among invitees, respondents, and collaborators, 88
D-2-2 Network science researchers network, 89
D-2-3 Researchers with high BC values and low BC values, 90
D-2-4 Largest component of the NSRN, 91
D-2-5 Disciplinary heterogeneity of the NSRN, 92

E-1   Schematic depiction of next-generation model for Army R&D showing the relationship between the main entities in this model, 99

TABLES

ES-1   Network Research Areas, 5

2-1   Representative Networks, 12
2-2   Maturity, Structure, Characteristics, and Impacts of Some Networks, 13

3-1   Network Research Areas, 24

8-1   Network Research Areas, 50

C-1   Representative List of Courses on Computer Science, 61
C-2   Real-World Networks Appearing in Courses, 62
C-3   Content of a Typical Network Science Course, 62
C-4   Network Models Commonly Used to Generate Network Topologies and Analytical Tools Used to Characterize and Study the Properties of Models, 63

D-1   Respondent's Country, 74
D-2   Canadian Respondent Provinces, 74
D-3   Respondent States, 76
D-4   Responses Per Field, 77
D-5   Respondent Affiliations, 78
D-6   Is Your Work Potentially Part of an Emerging Field of Network Science?, 78
D-7   Is There an Identifiable Field of Network Science?, 79
D-8   Summary Decomposition of the Input Attributes of Networks, 80
D-9   Summary Decomposition of the Derived Properties of Networks, 82
D-10 Summary Decomposition of Constraint Models, 83
D-11 Summary Decomposition of the Problem Dimensions of Networks, 83
D-12 Major Players and Cited Applications, 85
D-2-1 Researchers Who Are Frequently Mentioned and Listed as Collaborators, 88
D-2-2 Researchers Who Act as Gatekeepers, 89
D-2-3 Components in the NSRN, 90

BOXES

ES-1   Summary of Responses to the Statement of Task, 2

1-1   Network Science: Foundation of Our Connected Age, 8
1-2   Statement of Task, 8

2-1   Books Relevant to Network Science, 17

3-1   Case Studies in Net-centric Operations, 21
3-2   Dependence of Army Operations on Networks: An Example, 23

8-1   Summary of Responses to the Statement of Task, 47

D-1   NRC Network Science Survey, 66
D-2   Mapping the Social Network and Expertise of Network Science Researchers, 88

E-1   Case Study from the World Health Organization: Avian Influenza, 103

Acronyms and Abbreviations

AAAS	American Association for the Advancement of Science
AIDS	acquired immunodeficiency syndrome
APS	American Physical Society
BAST	Board on Army Science and Technology
BCT	brigade combat team
BFT	Blue Force Tracker
C3	command, control, and communications
C4ISR	command, control, communications, computers, intelligence, surveillance, and reconnaissance
CDC	Centers for Disease Control
COP	common operational picture
DARPA	Defense Advanced Research Projects Agency
DOD	Department of Defense
DOE	Department of Energy
FBCB2	Force XXI Battle Command Brigade and Below
FCS	Future Combat System
GIG	Global Information Grid
HIV	human immunodeficiency virus
IED	improvised explosive device
JCI	Joint Combat Identification
JNN	Joint Network Nodes
JTF	Joint Task Force
MAS	medical aid station
NAS	National Academy of Sciences
NASA	National Aeronautics and Space Administration
NCE	Networked Center of Excellence
NCO	network-centric operations
NCW	network-centric warfare

NIH	National Institutes of Health	
NRC	National Research Council	
NSF	National Science Foundation	
NSRN	network science researcher network	
NTC	National Training Center	
OCP	Open Control Platform	
OFT	Office of Force Transformation	
OPFOR	opposing force	
OSI	open system interconnection	
PI	principal investigator	
R&D	research and development	
RFP	Request for Proposal	
S&T	science and technology	
SARS	severe acute respiratory syndrome	
SBIR	Small Business Innovation Research	
SEC	Software-Enabled Control	
SIAM	Society for Industrial and Applied Mathematics	
TCP/IP	Transmission Control Protocol/Internet Protocol	
UAV	unmanned aerial vehicle	
UGC	unmanned ground vehicle	
URL	Uniform Resource Locator	
WHO	World Health Organization	
www	World Wide Web	

Executive Summary

Society depends on a diversity of complex networks for its very existence. In the physical sphere, these include the air transportation network, highways, railroads, the global shipping network, power grids, water distribution networks, supply networks, global financial networks, telephone systems, and the Internet. In the biological arena, they include genetic expression networks, metabolic networks, our bodies, ant colonies, herds, food webs, river basins, and the global ecological web of Earth itself. In the social domain, they include governments, businesses, universities, social clubs, churches, public and private school systems, and military organizations. The military's dependence on interacting networks in the physical, information, cognitive, and social domains is clear from its effort to transform itself into a force capable of network-centric operations (NCO).

In spite of society's profound dependence on networks, fundamental knowledge about them is primitive. Many physical networks—for example, global communication and transportation networks—have quite advanced technological implementations, but their behavior under stress still cannot be predicted reliably. For biological and social networks, scientists do not understand what they are, much less how they operate. There is a huge gap between what we need to know about networks to ensure the smooth working of society and the primitive state of our fundamental knowledge. This gap makes the military vision of NCO problematic, at best.

STUDY APPROACH

The present study was commissioned by the Army to find out whether identifying and funding a new field of investigation, "network science," could help close this gap. The chair worked with the NRC staff to nominate committee members representative of the broad scope of efforts in network research and also of the interests in this topic on the part of the Army.

At its initial meetings the committee focused on data collection tasks. Members were invited to present their ideas about the definition and content of network science. This exercise was expanded to encompass telephone interviews with a number of distinguished researchers and a questionnaire distributed inquiring about the role of networks in today's global economy and the military in particular. The committee also collected data on the use of networks in the military, learning from extensive reading and presentations at its second and third meetings. The results of these data-gathering tasks are reported in Chapters 2 through 4.

The committee formed two special task teams. One team surveyed academic courses on network research to determine the content of core knowledge about networks. The results of this effort are reported in Chapter 5 based on the data presented in Appendix C. The other task team developed and circulated the questionnaire to as broad a cross section of the network research community as possible given the time and financial constraints of this study. The committee's analysis of the responses is reported in Chapter 6 and Appendix D.

After characterizing the importance and content of network science, the committee turned its attention to the matter of how the Army might create value by investing in research on networks. This task was complicated by the fact that "the Army" is shorthand for a diverse group of constituencies with multiple agendas and priorities. The committee formed into new task teams to formulate three different investment scenarios that span the various interests and agendas. The scenarios are reported in Appendix E and are summarized, along with specific findings, in Chapter 7.

Representative literature used over the course of the study is listed in Appendix F. The body of the report—Chapters 2 through 7 and Appendixes C through E—contains the factual findings, and Chapter 8 contains the committee's conclusions and recommendations. Box ES-1 provides a summary of how the various report chapters respond to the statement of task.

1

> **BOX ES-1**
> **Summary of Responses to the Statement of Task**
>
> The Assistant Secretary of the Army (Acquisition, Logistics, and Technology) has requested the National Research Council (NRC) Board on Army Science and Technology (BAST) conduct a study to define the field of Network Science. The NRC will:
>
> 1. Determine whether initiation of a new field of investigation called Network Science would be appropriate to advance knowledge of complex systems and processes that exhibit network behaviors. If yes, how should it be defined?
>
> **A working definition of network science is the study of network representations of physical, biological, and social phenomena leading to predictive models of these phenomena. Initiation of a field of network science would be appropriate to provide a body of rigorous results that would improve the predictability of the engineering design of complex networks and also speed up basic research in a variety of applications areas (Chapter 4).**
>
> 2. Identify the fields that should comprise Network Science. What are the key research challenges necessary to enable progress in Network Science?
>
> **General consensus exists among practitioners of network research in diverse application areas on topics that constitute network science (Chapter 5). There are seven major research challenges (Chapter 6).**
>
> 3. Identify specific research issues and the theoretical, experimental, and practical challenges to advance the field of Network Science. Consider such things as facilities and equipment that might be needed. Determine investment priority, time frame for realization, and degree of commercial interest.
>
> **Current military concepts of "net-centricity" are based on applications of computer and information technology that are far removed from likely results of basic research in network science. Table ES-1 lists current areas of network research of interest to the Army, including priority, time frames, and commercial interest (Chapter 3).**
> **Current funding policies and priorities are unlikely to provide adequate fundamental knowledge about large complex networks that will advance network-centric operations. Besides the information domain, there are social, cognitive, and physical technology domains in the current conceptual framework for network-centric operations; there is no "biological" domain (Chapters 2–4).**
> **A basis for network science is perceived in different ways by the communities concerned with engineered, biological, and social networks at all levels of complexity. Basic research efforts are incoherent (Chapters 5 and 6).**
> **Options for obtaining value from investments in network science include scenarios ranging from building a base of basic research, to leveraging business practices for market-driven R&D in specific areas of network applications, to creating a robust capability for network-centric operations (Chapter 7).**
>
> 4. Given limited resources (and likely investments of others), recommend those relevant research areas that the Army should invest in to enable progress toward achieving Network-Centric Warfare capabilities.
>
> **Recommendations 1, 1a through 1d, 2, and 3 provide the Army with an actionable menu of alternatives that span the opportunities accessible to it. By selecting and implementing appropriate items from this menu, the Army can develop a robust network science to enable the desired progress (Chapter 8).**
>
> ---
> NOTE: The statement of task is in lightface; the summary of responses is in boldface.

OVERARCHING CONCLUSIONS

The committee reached three overarching conclusions about the significance of networks and the state of knowledge about them. First, it documented the pervasive influence of networks in all aspects of life—biological, physical, and social—and concluded that they are indispensable to the workings of the global economy and to the defense of the United States against both conventional military threats and the threat of terrorism.

Second, the fundamental knowledge needed to predict the properties of large infrastructure networks (such as the Internet and power grid) and vital social networks (the global economic system and military command and control) is primitive. Not even physical communications networks can be designed so that their resistance to failure and scaling up from small to large can be predicted a priori with confidence.

Networks are built on top of one another. Social networks, for example, are built on information networks, which in turn are built on communications networks that operate using physical networks for connectivity. The networks required to make NCO for the military a reality span the physical, information, cognitive, and social domains. They are interactive and mutually interdependent.

There is no science today that offers the fundamental knowledge necessary to design large, complex networks in such a way that their behaviors can be predicted prior to building them. Given this shortfall, trying to implement network-centric operations capabilities as envisioned by the Department of Defense (DOD) is like trying to design and build a modern combat jet aircraft without resorting to the science of fluid dynamics.

Third, in spite of the need for a science of networks and the high level of interest in the scientific community, current funding policies and practices of federal agencies are focused on specific network applications and are not focused on accumulating fundamental knowledge about networks.

Research on networks is fragmented. It is supported in disciplinary stovepipes that encourage jargon, parochial terms, and local values. Fundamentals of network structure, dynamics, and simulation are being rediscovered by different groups that emphasize uniqueness rather than a common intellectual heritage and methodologies. The fragmentation is aggravated by funding-agency policies and procedures that reward narrow disciplinary interests rather than results that are demonstrably usable for addressing national problems.

Nor is funding focused in areas with widespread application, such as the development of predictive models of social networks, which could directly impact vital national problems, from secondary education in urban slums to military command and control.

Although researchers, especially the best researchers, are reacting rationally to the incentives placed before them, these incentives reflect poorly the national interests of the United States in a globally connected world.

SPECIFIC CONCLUSIONS

On the basis of its data collection and analysis, the committee offers the following conclusions containing answers to the specific questions posed in the statement of task.

Different research communities give different answers to the question, What is network science? Nevertheless, the committee discerned some basic features. First, network science is distinct from both network technology and network research: It is characterized by the discovery mode of science rather than the invention mode of technology and engineering. Network research encompasses both.

Network science is broad in scope, encompassing physical, biological, and social networks. Synergies between network representations and models in these domains give it power. It creates fundamental knowledge that enables the a priori prediction of the behaviors of diverse networks in contrast to their a posteriori characterization. In short, network science consists of the study of network representations of physical, biological, and social phenomena, leading to predictive models of these phenomena.

The remarkable diversity and pervasiveness of network representations and models render network science a topic that can be leveraged by both civil society and the military. A provisional consensus exists around its core contents, making network science an identifiable area of investigation. Excellent research problems on a variety of topics exist. By making an investment in network science, the Army could forge a single approach to a diverse collection of applications. The committee therefore concludes that network science is an emerging field of investigation whose support would address important societal problems, including the Army's pursuit of NCO capabilities.

Although the boundaries of network science are fuzzy, the committee found broad consensus among practitioners in network applications—including physical, biological, social, and information networks—on the key topics, the types of tools that must be developed, and the research challenges that should be investigated. Based on the responses to its questionnaire and its own knowledge, the committee concluded that there are seven major research challenges, the surmounting of which will enable progress in network science:

- *Dynamics, spatial location, and information propagation in networks.* Better understanding of the relationship between the architecture of a network and its function is needed.
- *Modeling and analysis of very large networks.* Tools, abstractions, and approximations are needed that allow reasoning about large-scale networks, as well as techniques for modeling networks characterized by noisy and incomplete data.
- *Design and synthesis of networks.* Techniques are needed to design or modify a network to obtain desired properties.
- *Increasing the level of rigor and mathematical structure.* Many of the respondents to the questionnaire felt that the current state of the art in network science did not have an appropriately rigorous mathematical basis.
- *Abstracting common concepts across fields.* The disparate disciplines need common concepts defined across network science.
- *Better experiments and measurements of network structure.* Current data sets on large-scale networks tend to be sparse, and tools for investigating their structure and function are limited.
- *Robustness and security of networks.* Finally, there is a clear need to better understand and design networked

systems that are both robust to variations in the components (including localized failures) and secure against hostile intent.

Finally, although all the military services have a vision of the future in which engineered communications networks play a fundamental role, there is no methodology for ensuring that these networks are scalable, reliable, robust, and secure. Of particular importance is the ability to design networks whose behaviors are predictable in their intended domains of applications. This also is true in the commercial sphere. The committee therefore concluded that the high value attached to the efficient and failure-free operation of global engineered networks makes their design, scaling, and operation a national priority.

The ultimate value derived from these engineered networks depends on the effectiveness with which humans use them. These uses can be beneficial (e.g., better combat effectiveness) or detrimental (e.g., their exploitation by criminal and terrorist groups). Therefore research into the interaction of social and engineered networks is also a national priority.

RECOMMENDATIONS

The statement of task asks the committee to recommend investments the Army should make in network science. The impact of networks on society transcends their impact on military applications, although both are vital aspects of the total picture. The current state of knowledge about networks is insufficient to support the design and operation of complex global networks for military, political, and economic applications. Advances in network science are therefore essential for developing adequate knowledge for these applications.

Recommendation 1. The federal government should initiate a focused program of research and development to close the gap between currently available knowledge about networks and the knowledge required to characterize and sustain the complex global networks on which the well-being of the United States has come to depend.

This recommendation is buttressed by increasing evidence that disruptive social networks (e.g., terrorists, criminals) learn to exploit evolving infrastructure networks (e.g., communications or transportation) in ways that the creators of these networks did not anticipate. The global war on terrorism, which is a main driver of military transformation, is only one recent manifestation of this general pattern. Addressing problems resulting from the interaction of social and engineered networks is an example of a compelling national issue that transcends the transformation of the military and that is largely untouched by current research on networks.

Within this broad context, recommendations 1a, 1b, and 1c provide the Army with three options:

Recommendation 1a. The Army, in coordination with other federal agencies, should underwrite a broad network research initiative that includes substantial resources for both military and nonmilitary applications that would address military, economic, criminal, and terrorist threats.

The Army can lead the country in creating a base of network science knowledge that can support applications for both the Army and the country at large. Maximum impact could be obtained by a coordinated effort across a variety of federal agencies, including DOD and the Department of Homeland Security, to create a national program of R&D focused on network science to develop applications that support not only network-centric operations but also countermeasures against international terrorist and criminal threats.

Alternatively, if the Army is restricted to working just with the DOD, it should initiate a focused program to create NCO capabilities across all the services.

Recommendation 1b. If the Army wants to exploit fully applications in the information domain for military operations in a reasonable time frame and at an affordable cost, it should champion the initiation of a high-priority, focused DOD effort to create a realizable vision of the associated capabilities and to lay out a trajectory for its realization.

Finally, if the Army elects to apply the insight from the committee primarily to its own operations, then it can still provide leadership in network science research.

Recommendation 1c. The Army should support an aggressive program of both basic and applied research to improve its NCO capabilities.

Specific areas of research of interest to the Army are shown in Table ES-1. This table expresses the committee's assessment of the relative priorities for these areas, the time frames in which one might reasonably expect them to be consummated as actionable technology investment options, and the degree of commercial interest in exploiting promising options. The committee notes that both trained personnel and promising research problems exist in many of these areas, so that the Army should be able to create productive programs readily.

Regardless of which options are adopted, however, Army initiatives in network science should be grounded in basic research as follows:

Recommendation 1d. The initiatives recommended in 1, 1a, 1b, and 1c should include not only theoretical studies

TABLE ES-1 Network Research Areas

Research Area	Key Objective	Time Frame	Commercial Interest	Priority for Army Investment
Modeling, simulating, testing, and prototyping very large networks	Practical deployment tool sets	Mid term	High	High
Command and control of joint/combined networked forces	Networked properties of connected heterogeneous systems	Mid term	Medium	High
Impact of network structure on organizational behavior	Dynamics of networked organizational behavior	Mid term	Medium	High
Security and information assurance of networks	Properties of networks that enhance survival	Near term	High	High
Relationship of network structure to scalability and reliability	Characteristics of robust or dominant networks	Mid term	Medium	Medium
Managing network complexity	Properties of networks that promote simplicity and connectivity	Near term	High	High
Improving shared situational awareness of networked elements	Self-synchronization of networks	Mid term	Medium	High
Enhanced network-centric mission effectiveness	Individual and organizational training designs	Far term	Medium	Medium
Advanced network-based sensor fusion	Impact of control systems theory	Mid term	High	Medium
Hunter-prey relationships	Algorithms and models for adversary behaviors	Mid term	Low	High
Swarming behavior	Self-organizing UAV/UGV; self-healing	Mid term	Low	Medium
Metabolic and gene expression networks	Soldier performance enhancement	Near term	Medium	Medium

but also the experimental testing of new ideas in settings sufficiently realistic to verify or disprove their use for intended applications.

By selecting from Recommendations 1a through 1c an option that is ambitious yet achievable, the Army can lead the country in creating a base of knowledge about network science that is adequate to support applications on which both the Army and the country at large depend.

The Army has another investment scenario that it could pursue: "building the base" for network science by funding a small program of basic research in network science. This investment of small amounts of Army risk capital funds would create a base of knowledge and personnel from which the Army could launch an attack on practical problems that arise as it tries to provide NCO capabilities.

If the Army is limited to modest changes in the funding of its R&D portfolio and incremental changes to the way that it manages these investments, funding only a small program of basic research in network science could still have a significant effect. But the committee wants to be crystal clear that investments in basic (6.1) research in network science have no immediate prospects of impacting the design, test, evaluation, and sourcing of NCO capabilities.

The main values created by a basic research investment would include access to thought leaders (principal investigators) in the university community, the training of students through their work on university projects, the development of a community that the Army can access to address its practical problems, and the efficient use of research dollars to impact multiple areas of application. To exploit these opportunities, the committee offers the following two recommendations:

Recommendation 2. The Army should make a modest investment of at least $10 million per year to support a diverse portfolio of basic (6.1) network research that promises high leverage for the dollars invested and is clearly different from existing investments by other federal agencies like the National Science Foundation (NSF), the Department of Energy (DOE), and the National Institutes of Health (NIH).

This modest level of investment is compatible with the Army's current R&D portfolio. There is an adequate supply of promising research topics and talented researchers to make this investment productive. Additionally, it can be implemented within the Army's current R&D management work processes.

To identify the topics in basic network science research that would bring the most value to NCO, the committee recalls that the open system architectures for computer networks consist of layers, each of which performs a special function regarded as a "service" by the layers above. It is useful to distinguish among the lower (physical and transport) layers of this architecture, the higher (applications) layers that are built on top of them to offer services to people, and the cognitive and social networks that are built higher still, on top of the services-to-humans layers.

Research on the lower layers of the network architecture is relatively mature. Improving these levels is more of an engineering problem than one requiring basic research. The most immediate payoffs from network science are likely to result from research associated with the upper levels of the network architecture and the social networks that are built at an even higher level. This is where the committee thinks that Army investments are most likely to create the greatest value.

An area of particular promise that has little or no current investment is the social implications of NCO for the organizational structure and command and control. Basic research could provide valuable insight into how military personnel use advanced information exchange capabilities to improve combat effectiveness. For example, one might study how troops in combat could use these capabilities to make better decisions. Additional basic research in the core content of network science might help to determine how the Army can most productively utilize the capabilities of its advanced information infrastructure.

Recommendation 3. The Army should fund a basic research program to explore the interaction between information networks and the social networks that utilize them.

The Army can implement Recommendations 2 and 3 within the confines of its present policies and procedures. They require neither substantial replanning nor the orchestration of joint Army/university/industry research projects. They create significant value and are actionable immediately.

The committee's Recommendations 1, 1a through 1d, 2, and 3 give the Army an actionable menu of options that span the opportunity space available. By selecting and implementing appropriate items from this menu, the Army can develop a robust network science to "enable progress toward achieving Network Centric Warfare capabilities," as requested in the statement of task.

1

Introduction

Network effects are found in biologically diverse worlds, at many layers of abstraction from micro to macro. These include molecular biochemical reactions, cellular neural networks, insect swarms, and entire ecologies. They also are found in such diverse engineered systems as power grids, communications networks, like the Internet, and the transportation infrastructure. Network effects are, however, most commonly associated with human social structures—we speak about networking as an essential skill for both doing our jobs and getting new ones. This dimension of networks has taken on special significance in the past few years as we recognize the powerful influence on society of criminal and terrorist social networks that exploit modern communication and transportation networks (Arquilla and Ronfeldt, 2001).

In the military, network effects occur in the communication systems that link platforms and Soldiers. The concept of network-centric warfare (NCW) takes the importance of networks for the military even further. In this concept dynamic battlefield command and control networks are built in real time, relying on more static networks such as physical communications, weapons systems platforms, and military organizational structure. The differences between static and dynamic networks are, however, not clearly understood, and our understanding of dynamic network effects is primitive.

Networks also build upon each other in layers—for example, a network of business process applications is built on a communications network that is, in turn, built on a physical network.

Despite the tremendous variety of complex networks in the natural, physical, and social worlds, little is known scientifically about the common rules that underlie all networks. This is even truer for interacting networks. Ideas put forth by scientists, technologists, and researchers in a wide variety of fields have been coalescing over the past decade, creating a new field of thinking—the science of networks (see Box 1-1).

Does a science of networks exist? Opinions differ. But if it does, network science is in its infancy and still needs to demonstrate its soundness as a science on which to base useful applications. The purposes of this report are to assess the scope and content of network science and to envision how its pursuit can create value for the United States in general and for the U.S. Army in particular. Semantics aside, the opportunity is historic. The unrelenting drop in the costs of computing, symbolized by Moore's law, has made massive computation a commodity, cheap and available to industry and individuals alike (Rheingold, 2002).

The benefits of connectivity—as quantified, for instance, by Metcalfe's law and first recognized in the rail and telephone networks—make it irresistibly attractive (Rheingold, 2002). The value of community—asserted, for example, in Reed's law—elevates connectivity to an economic imperative (Rheingold, 2002). The linkage of information networks has led to a global information grid, to which the majority of the world's population is likely to be connected within the next decade. This situation is unparalleled in human history. It will lead to social institutions and human behaviors never before seen or anticipated (Ronfeldt, 2005). Its initial consequences already are being reported in the popular press (Business Week, 2005). It renders the study of networks and their effects—the pursuit of network science—a social, scientific, and technological imperative for the 21st century.

Why should the military in general and the Army in particular care? Aside from the fact that the U.S. military is embedded in this wave of technological and social change, the exploration of network science promises insights and tools that are indispensable to improving its combat effectiveness in the new world of likely conflicts.

The development of the Army's Future Combat Systems (FCS) is experiencing cost and schedule overruns because of the immense complexity of the effort (Weiner, 2005). Given the committee's findings about the immaturity of network science, this is hardly surprising. Designing and testing the FCS communications network alone is like trying to design and test a modern jet aircraft without the benefit of the science of aerodynamics or like designing and testing a radio or

> **BOX 1-1**
> **Network Science: Foundation of Our Connected Age**
>
> The first thing you see in the room on your right as you enter Boston's Museum of Science is a vertical peg board about 8 feet square. Sticking out of a checkerboard square pattern are pegs at the corner of every square. Ping-pong-size balls drop from the center top of the board and carom crazily off various pegs on the way down to the bottom. There is no way to predict which way a ball will zig or zag at each row or where any one ball will land in the bottom row. The pathway of each drop is completely random. Yet, despite the chaos at the beginning, the balls collect at the bottom in a perfect normal curve distribution. Within moments, the system changes from totally random individual actions to a completely symmetrical and predictable aggregate—order emerges out of chaos. Similar phenomena are exhibited by human social networks, like the complex web of traders and investors on the New York Stock Exchange (Bernstein, 1992) or the operation of any large city (Johnson, 2001; Watts, 2003). Why? Such is the mystery of self-organization in large, complex networks.
>
> Similarly, what accounts for how birds flock and fish school? Why do accumulated grains of sand build to a mound or dune until finally one grain proves to be a grain too many and an avalanche occurs? And why do electrical systems crash when they reach comparable tipping points? Why are we all connected by the famous six degrees of separation? What do epidemics, earthquakes, computer viruses, religious fundamentalism, and the "Friends of Kevin Bacon" game all have in common?
>
> The common element in the answers to these questions is that things are connected. Connections create networks, networks operate by rules and probably laws, and a new science of networks is emerging to determine and explain what these are. Nations, species, corporations, and armies will all be affected by this new science.

TV without the benefit of the fundamental knowledge of electromagnetic waves.

The engineering of complex physical networks, like that of the FCS, is not predictable because the scientific basis for constructing and evaluating such designs is immature. This is even more the case for characterizing, modeling, and evaluating modern criminal and terrorist networks that are built on a physical communications network infrastructure (Arquilla and Ronfeldt, 2001). The Office of Force Transformation (OFT) has advanced the concepts of network-centric warfare (Cebrowski and Garstka, 1998) and network-centric operations (Garstka and Alberts, 2004) to define warfare in the 21st century. Both concepts involve multiple interacting networks built one on top of the other. Neither has a firm empirical and analytical base. Thus, getting a grip on the fundamental science of networks—their structure and dynamics—is a topic of pressing concern for military as well as political and economic interests.

SCOPE OF THE STUDY

Recognizing the urgency of this situation, the National Research Council (NRC) Board on Army Science and Technology (BAST) formed the Committee on Network Science for Future Army Applications. The statement of task for the committee consists of four charges (Box 1-2). This document is the report of that committee.

STUDY APPROACH AND CONSTRAINTS

Special care was devoted to the composition of the committee. Biographies are given in Appendix A. Three representative groups of members were selected. The first group included individuals from the physical sciences, engineering, biological sciences, and social sciences research communities. In order to sample the breadth of intellectual effort on network science, committee members were selected who have recent first-hand experience in the subject matter as reflected in their recent books or research and teaching assignments. Thus, committee membership includes the authors of *Six Degrees: The Science of the Connected Age*

> **BOX 1-2**
> **Statement of Task**
>
> The Assistant Secretary of the Army (Acquisition, Logistics, and Technology) has requested the National Research Council (NRC) Board on Army Science and Technology (BAST) conduct a study to define the field of Network Science. The NRC will:
>
> - Determine whether initiation of a new field of investigation called Network Science would be appropriate to advance knowledge of complex systems and processes that exhibit network behaviors. If yes, how should it be defined?
> - Identify the fields that should comprise Network Science. What are the key research challenges necessary to enable progress in Network Science?
> - Identify specific research issues and the theoretical, experimental, and practical challenges to advance the field of Network Science. Consider such things as facilities and equipment that might be needed. Determine investment priority, time frame for realization, and degree of commercial interest.
> - Given limited resources (and likely investments of others), recommend those relevant research areas that the Army should invest in to enable progress toward achieving Network-Centric Warfare capabilities.

(Duncan Watts); *Linked: The New Science of Networks* (Albert-Laszlo Barabási); *The Future of Work* (Thomas Malone); and *It's Alive: The Coming Convergence of Information, Biology and Business* (Stan Davis); and the editor of *Control in an Information Rich World: Report of the Panel on Future Directions in Control, Dynamics and Systems* (Richard M. Murray), a report of the Society for Industrial and Applied Mathematics (SIAM). The committee also includes the organizer of the new systems biology curriculum at Harvard University (Pamela A. Silver) and a contributor to the World Technology Evaluation Center's assessment of systems biology (Adam Arkin). Thus, active players in the diverse communities engaged in creating the science of modern networks are represented on the committee.

The second representative group comprises both the command and the research and development (R&D) communities of the military, including retired flag officers and experts with experience at the Defense Advanced Research Projects Agency (DARPA) (Ronald J. Brachman), the National Defense University (Robert E. Armstrong), and MITRE (Norval L. Broome). Paul Van Riper, who first applied the concepts of network science to the articulation of the command-and-control doctrine in the U.S. Marine Corps, is among the flag officers, as are William Hilsman, former chief information officer of the Army and Jack Pellicci, a retired Army brigadier general who is now an executive for the Oracle Corporation.

The third group includes representatives from the academic and industrial management worlds. Two deans (Richard DeMillo, Georgia Institute of Technology, and John E. Hopcroft, Cornell) played a major role in defining the scope of the study. Dr. DeMillo also has served as chief technology officer of Hewlett-Packard. The committee's outreach efforts were led by Richard Murray (California Institute of Technology) and by Will E. Leland, chief scientist, Telcordia Technologies. The committee chair Charles Duke, has been an R&D manager for 22 years at Xerox and was chief scientist and deputy director of the Pacific Northwest National Laboratory.

Clearly, the committee membership spans the diverse constituencies of network science: military commanders, business managers, program managers, research managers, and active researchers. Together, the members had the skills and experience needed to assess the content of network science; its prospects for advancing engineering, social, and biological technologies; and the potential for selected research efforts to impact the U.S. military in general and the U.S. Army in particular over different timescales.

Initially the committee was divided into three working teams. Team I devoted its attention to assessing the impacts of past network science and technology and to extrapolating this record to project future impacts. Team II focused on defining the scope of network science. Its members identified the core elements of network science underpinning the diverse array of applications and technologies in the social, economic, engineering, and biological arenas. Team III concentrated on outreach to communities that currently practice network science and technology. It identified community members via literature studies, interviews, and e-mail inquiries. It constructed and circulated a Web-based questionnaire. From the responses, it extracted the recognized core content of network science, research activities in which community members are engaged, and their perception of the major research challenges.

The three streams of activity carried out by the teams were brought together midway though the committee's deliberations with the writing of a full-message draft. Consensus was reached on the findings pertinent to charges (1) and (2) in the statement of task. Then, in response to charges (3) and (4) the committee was reconstituted into three new teams, which developed scenarios of how network science could add value for the Army. The committee's findings, conclusions, and recommendations were then refined and ratified at the final meeting.

REPORT ORGANIZATION

This report documents the study approach, findings, conclusions, and recommendations. It is organized in accord with the statement of task in Box 1-2 and the study approach described above. Where the committee conducted research and discovered factual information, the information is reported as a finding. Multiple findings combine as the basis for conclusions, some "overarching" and the rest "specific" conclusions pertinent to specific requests in the statement of task. The conclusions are contained in Chapter 8 along with the committee's recommendations.

The bulk of the committee's results are reported as findings in Chapters 2 through 7. Some of the details that support these findings are presented in the appendixes; others may be found in the references cited in the text.

Chapter 2 characterizes the pervasive impact of networks and network research in the 21st century. Chapter 3 describes their significance for the military in general and the Army in particular. Chapter 4 offers a provisional definition of "network science" and notes the promise afforded by developing a science of networks. Chapter 5 describes the potential scope and content of network science, as determined from an analysis of courses at academic institutions worldwide. The contents of these courses are indicated in Appendix C. Chapter 6 discusses the current status of network science and identifies associated research challenges. This material is based on an analysis, presented in Appendix D, of the results of a questionnaire sent to over 1,000 researchers working on various topics pertaining to networks. Chapter 7 presents findings concerning how the Army can create value from investments in network science. It is based on the investment scenarios presented in Appendix E. Finally, as noted above, Chapter 8 contains the committee's conclusions and recommendations.

REFERENCES

Arquilla, J., and D. Ronfeldt. 2001. Networks and Netwars: The Future of Terror, Crime and Militancy. Santa Monica, Calif.: RAND.

Bernstein, P.J. 1992. Capital Ideas: The Improbable Origins of Modern Wall Street. New York, N.Y.: Free Press.

Business Week. 2005. The power of us. Pp. 74–82.

Cebrowski, A., and J. Garstka. 1998. Network centric warfare. Proceedings of the United States Naval Institute 24: 28–35.

Garstka, J., and D. Alberts. 2004. Network Centric Operations Conceptual Framework Version 2.0. Vienna, Va.: Evidence Based Research, Inc.

Johnson, S. 2001. Emergence: The Connected Lives of Ants, Brains, Cities and Software. New York, N.Y.: Scribner Associates, Inc.

Rheingold, H. 2002. Smart Mobs: The Next Social Revolution. Cambridge, Mass.: MA Basic Books.

Ronfeldt, D. 2005. A long look ahead: NGOs, networks, and future social evolution. In Environmentalism and the Technologies of Tomorrow, R. Olson and D. Rejeski, eds. Washington, D.C.: Island Press.

Watts, D.J. 2003. Six Degrees: The Science of a Connected Age. New York, N.Y.: W.W. Norton.

Weiner, T. 2005. Drive to build high-tech army hits cost snags. New York Times, March 28, 2005.

2

Networks and Network Research in the 21st Century

If there is one word to describe society in the early 21st century, it surely must be "connected." We have grown up taking for granted the vast interlinked networks that bring electricity, water, gas, and cable TV to our homes and that allow us to be in personal contact with others almost anywhere in the world by telephone, e-mail, and other communications means. The Internet, especially the World Wide Web (www), is thoroughly ingrained in our everyday lives. The defense of our nation is heavily dependent on electronic networks for communication, command and control, collaborative decision making, intelligence gathering, and other critical functions.

While less obviously "networks," many other richly connected systems play crucial roles in our lives. When diseases are transmitted by person-to-person contact, their spread patterns and ultimate effect are highly dependent on connections that can be described as a network. When cells divide and transform under the influence of minute amounts of biochemical elements in the body, they trigger a network of influences and dependent reactions. Human organizations are networks, often captured graphically with organization charts. In our daily lives we encounter health-care provider networks, purchase goods from companies that acquired them from supply networks, and pay for them using networks of banks and credit card companies. Our brains are immense networks of highly interconnected nerve cells, responsible for our ability to see and hear, make decisions, remember and learn, and act.

In order to get a sense of the scope and character of these networks, the committee classified them into biological (e.g., metabolic pathways), physical (e.g., the power grid and telephone system), and social (e.g., governments and churches). This taxonomy is developed in Table 2-1, which identifies some important physical, social, and biological networks and gives an indication of their global impact. This table illustrates clearly the utter pervasiveness of networks in every aspect not only of human existence but also of the existence of all living entities on planet Earth. Connectivity is an essential ingredient of life as we know it.

Not only are networks pervasive, they are astonishingly diverse. Moreover, they can be characterized by figures of merit that indicate how large, how complicated, how robust, and how important they are. This aspect of networks is illustrated in Table 2-2, which shows the characteristics of a diverse sampling of networks. The figures of merit for the columns are defined in the footnotes. This table is worthy of close examination because it reveals the wide diversity of the scales, structures, states of maturity, technological intensity, benefits, and consequences of failure of some of the networks that we encounter daily. It also illustrates how complex some of the networks are, leading one to wonder if their designers and operators can control their behaviors.

Inspection of Tables 2-1 and 2-2 leads to the committee's first finding:

Finding 2-1. Networks enable the necessities and conveniences of modern life.

The tables illustrate how vital networks are to modern life. We see from them that networks underlie nearly every aspect of the infrastructure that supports daily life. Electricity, water, transportation, telephone service, Internet connection, health care, banking, shopping, education, and government all are brought to us by physical or social networks.

Our bodies and minds are also manifestations of networks. The natural world in which we live is a vast array of ecological networks. Networks are ubiquitous in daily life. They also are central to the global economic infrastructure. The failure of any of these networks impacts society.

Finding 2-2. Engineered networks are a major driver of the increasingly global economy and can be of benefit to both the United States and its competitors.

It can be seen from these tables that modern communications and transportation networks are the drivers of the global economy. They provide the fundamental connectivity on which global banking, product design, tourism, supply chain

TABLE 2-1 Representative Networks

Biological Networks[a]		Physical Networks		Social Networks	
Type of Network	Global Impact	Type of Network	Global Impact	Type of Network	Global Impact
Disease transmitting networks (HIV, influenza, TB, malaria, cholera)	Spread of disease, epidemics	Distribution grids (electric power, water supply, business supply chains)	Efficient distribution of goods or commodities	Affiliation/ acquaintance networks (terrorist, community, business, religious, clubs)	Efficient collaboration and activity coordination
Ecological networks (food webs, river basins, rain forest)	Survival of selected species; global weather and topography	Telecommunications infrastructure (cellular, PSTN, cable TV, Internet)	Instantaneous worldwide information distribution	Broadcast networks (radio, TV networks like NBC, CBS, CNN)	Dissemination of identical information to large groups
Metabolic networks	Sustenance of life for a given generation of living entities	DOD global information grid (sensors, communications, and weapons)	Network-centric warfare and network-enabled operations	Information exchange networks (U.S. mail, local and long-distance telephone service)	Cheap, convenient long distance pair-wise communications
Community networks (insect societies, animal herds, bird flocks, schools of fish)	Survival of selected species	Transportation networks (airports, highways, railways, shipping)	Rapid movement of goods from supplier to market; modern travel	Group forming networks (eBay, corporate intranets)	Easy, convenient formation of groups of like-minded people who have never met
Gene expression networks	Transmission and evolution of life between generations	Electronic financial transaction networks (banking, credit cards, ATMs)	Electronic cashless transactions	Supply chains and business networks	Coordination of multiple players to achieve common goals, global cost reduction
				Social services networks (Social Security, family services, Medicare, Medicaid)	Efficient delivery of government services to large, distributed constituencies

NOTE: PSTN, public switched telephone network; DOD, Department of Defense.
[a]Includes biochemical and other networks that are natural rather than manmade.

management, and customer relationship management depend. From overseas manufacturer to Wal-Mart, products are ordered electronically, produced on demand, shipped around the globe by shipping networks, and delivered to local stores by rail and truck networks. Heralded by some pundits as the century of biology and nanotechnology, the 21st century is in fact an era of networks and is called by others "The Age of Information and Telecommunications" (Perez, 2002).

Finding 2-3. Social and biological networks bear important similarities to engineered networks.

One might infer from Table 2-1 that biological and social networks are similar to engineered networks. Indeed, much recent literature has been devoted to documenting just how this is the case (Barabási, 2002; Bower and Bolouri, 2001; Dorogovtsev and Mendes, 2003; Newman, 2003; Watts, 2003). The important point for this study is that many methods and models are applicable to networks of all kinds: biological, physical, and social. (These commonalities are explored in Chapter 5 and Appendix C.)

Finding 2-4. Advances in computer-based technologies and telecommunications are enabling social networks that facilitate group affiliations, including terrorist networks.

An important property of networks is that they may be built on top of each other. For example, a social network may be formed based on an information network built on a communications network that utilizes a physical network of transmission equipment. This property enables experimentation in the social network realm using commercial com-

TABLE 2-2 Maturity, Structure, Characteristics, and Impacts of Some Networks

Sample Network	Relative Maturity[a] (High, Medium, Low)	Network Structure[b]			Technology Intensity[c] (High, Medium, Low)	Network Scope[d]	Representative Societal Impacts/Benefits	Societal Impact of Catastrophic Failure[e] (High, Low)	Catastrophic Failure Description
		Number of Nodes	Topology Complexity	Scaling					
U.S. electric power distribution grid	High	High	Low	N	Medium	National	Electric lighting, appliances, and electronics	High	Continent-spanning blackout
Air transportation network	High	Medium	Medium	$N^{**}2$	High	Regional/ national/ global	Rapid global transport of people and cargo	Low	Major weather-related delays
Integrated circuits (chip level)	Medium	Medium	Medium	$N^{**}2$	High	Local	Ubiquitous computing and other electronic devices	Low	Device failure or recall
Cellular network and public switched telephone network	High	High	Low	N	High	National/ global	Instantaneous mobile worldwide communications	High	Surge-caused outage during a crisis
Sexual networks (e.g., those leading to or spreading HIV or sexual diseases)	N/A	Low	Low	N	N/A	Mostly local, but with modern transportation can be regional, national, or global	Large segments of population afflicted with AIDS in underdeveloped world	High	Onset of global pandemic
Internet data-link layer (router) topology	High	Medium	High	$N^{**}2$	High	Global	Enabler of Web and electronic commerce	Low	Major denial-of-service attacks
Applications layer Internet topology	Medium	Medium	Medium	$2^{**}N$	High	Global	Support for group-forming networks	Low	Computer viruses, spyware, and identification theft
Bank of America financial and banking network	Medium	Medium	Low	$N^{**}2$	High	National/ global	Cashless retailing and electronic currency exchanges	High	Global disruption of electronic financial transactions
Wal-Mart-like business supply chain	Low	Low	Medium	$2^{**}N$	Medium	National/ global	Just-in-time supply and inventory control	Low	Stock items not in stores
Small (50,000 or less) town governments	High	Low	Medium	$2^{**}N$	Medium	Local	Roads, water, sewage, zoning, police	Low (as individual governments)	Loss of local order, e.g., looting

[a]The network's position on a scale starting from first generation (at emergence, a low state) and ending with a high state of maturity, by which time the network has gone through multiple subsequent iterations.
[b]Network structure is characterized by number of nodes, topology complexity, and scaling. The number of nodes ranges between low (<1,000), medium (1,000 to 10,000,000), and high (>10,000,000). Topology complexity describes the diversity of interconnections from varied and complex to simple and

continues

TABLE 2-2 Continued

uniform. Scaling means economic or social value of that network as a function of N (the number of nodes). A linear value of N means that service is aimed at individual users. $N**2$ is the value that results from person-to-person transactions, and $2**N$, the value that results from the establishment of group affiliations (Rheingold, 2002, p. 58).

^{c}Network topology that is enabled by or highly dependent on modern computer-to-computer communications technologies. The high, medium, and low ranges are determined by the approximate number of computers in the network—for example, high range indicates >10^6 computers in the network, medium is 10^6 to 10^3, and low is <10^3.

^{d}Geographical scope of the network: global, national, regional, or local.

^{e}Potential consequences for society at large of a failure that is extremely destructive yet highly improbable. High range means >$100 million and low means <$100 million.

munications and information networks. The implications of this fact for criminal, terror, protest, and insurgency networks have been explored by Arquilla and Ronfeldt (2001) and are a common topic of discussion by groups like the Highlands Forum, which perceive that the United States is highly vulnerable to the interruption of critical networks.

Finding 2-5. The high value attached to common engineered networks makes their design, scaling, and operation a topic of national priority.

It seems self-evident from the ubiquity of networks in our daily lives, as well as from the potential for massive civil disruption if they fail, that the design, scaling, and operation of networks is a national priority. Failures and threats of failure caused by terrorist action add urgency to the strategic imperative to design, deploy, and operate robust networks whose behaviors are predictable. Yet as we shall see later in this report, fundamental (as opposed to empirical) knowledge about how to do this is primitive. The current state of knowledge about network design and characterization is roughly analogous to the state of knowledge about metallurgy in Europe in the 16th century. The empirical steel-forming technology of the day was sufficiently advanced to enable Europe to conquer most of the world but provided only a pale indication of the materials designs that would become possible in the 20th century based on the science of metallurgy (Diamond, 1999). As the committee looked at the current state of network technology and research in relation to our near total dependence on networks, it was amazed at the abundance of interest in network applications and the lack of fundamental scientific research that might advance the development of an underlying network science to support the study of networks in general.

Finding 2-6. Interest in network research has exploded during the past 5 years.

Research on networks has become highly visible during the past 5 years. A number of the measures that normally accompany the emergence of a new field document this heightened interest. For example, the number of publications focusing on complex networks increased significantly during that time. A "complex network" is one that exhibits emergent behaviors that cannot be predicted a priori from known properties of the network's constituents (Boccara, 2004). This is not limited to a single discipline, but it is stimulated by simultaneous interest in communication systems (like the Internet or the Web), biological systems (like metabolic or protein interaction networks), and social systems (like collaboration or e-mail networks). As shown in Figure 2-1, the number of publications with "complex networks" in their title has increased fourfold over the past 5 years without signs of saturation. The two most cited papers on complex networks have been cited more than 1,500 times, according to Google Scholar (Watts and Strogatz, 1998; Barabási and Albert, 1999). Fueled by this increasing interest, major scientific journals have devoted special issues, reviews, or editorials to the promise of networks. For example, *Nature* has published several reviews on the subject (Strogatz, 2001; Ottino, 2004; Koonin et al., 2002). *Science* and the *Proceedings of the National Academy of Sciences* devoted special issues to it (Jasny and Ray, 2003; PNAS, 2004). Two major physics review journals, *Reviews of Modern Physics* and *Advances in Physics*, have each published highly cited reviews on networks (Albert and Barabási, 2002; Dorogovtsev and Mendes, 2002). So have major engineering, applied mathematics, and sociology journals, like *IEEE Control Systems Magazine*, *SIAM Review*, and *Annual Review of Sociology* (Amin, 2002; Watts, 2004; Newman, 2003). Furthermore, several of the most prominent biology journals, like *Nature Reviews Genetics*, *Nature Immunology*, *Nature Structural Biology*, and *Nature Reviews Systems Biology*, have published high-profile reviews and editorials, often highlighting them on the cover of the journal, as shown in Figure 2-2.

Another sign of the emergence of a network science community is the organization of meetings devoted to network research. Indeed, during the past 5 years more than 20 international conferences and workshops and summer schools have focused exclusively on network research, some drawing close to 400 participants. Electrical engineering and computer science conferences have for many years devoted entire sessions to networks. In addition, major physics and biology meetings devote many focus sessions to networks. For example, the March 2004 and March 2005 meetings of

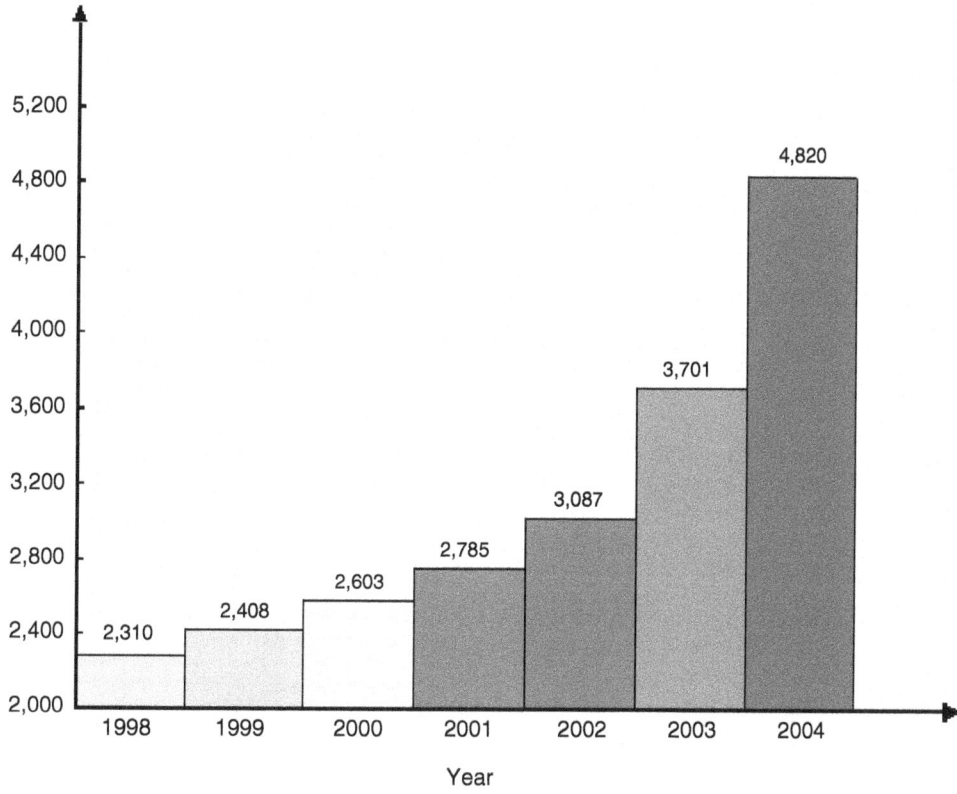

FIGURE 2-1 Number of papers with the term "complex network" in the title. SOURCE: Introduction to Complex Networks: Modeling, Control and Synchronization. Briefing by Guanrong Chen, director, Centre for Chaos Control and Synchronization, City University of Hong Kong, to the First Chinese Conference on Complex Networks, Wuhan, China, April 2005.

the American Physical Society (APS), the largest annual physics meeting in the world, had more than 10 network-related sessions. Similarly, the American Association for the Advancement of Science (AAAS) regularly features symposia devoted to network science and its applications.

The number of books on network science has exploded. Four general-audience books, translated into over 10 languages and making the bestseller list in several countries, introduced the promise of network science to the general public (Barabási, 2002; Buchanan, 2002; Watts, 2003; Huberman, 2001). Box 2-1 also lists over 15 monographs focusing either on network science in general or on its applications to specific fields.

In 2004, responding to the public's fascination with networks, the New York Hall of Science opened a major exhibit entitled "Connections: Seeing the World in a Different Way." The exhibit focused on the impact of networks on science, technology, and the arts.

Finally, as is discussed in Chapter 5 and Appendix C, most major U.S. universities have developed courses on various aspects of network science. These are offered in a variety of departments, including electrical engineering, physics, computer science, biology, economics, and sociology.

The European Union recognized the potential of research in networks early on under its Sixth Framework Program[1] by investing several million euros per year in flagship programs, such as COSIN, EVERGROW, DELIS, and EXYSTENCE, which focus on complex networks and their applications (Amaral et al., 2004).

In addition to providing the knowledge underlying the design and operation of many of the global communications, transportation, and power infrastructures noted in Tables 2-1 and 2-2, network research is leading to the creation of new businesses. The poster child of the benefits of network thinking is Google and all the second-generation search engines built after Google. Indeed, Google's phenomenal success and what sets it apart from its early competitors is its revolutionary algorithm, which used the topology of the Web to rank the obtained search results. Google is a wonderful example of how a piece of published research by two graduate students on random walks on networks, an academic exercise,

---

[1]For further information, see http://www.cordis.lu/ist/fet/co.htm. Accessed August 19, 2005.

FIGURE 2-2 Magazines and journals with articles on networks.

led in less than a decade to a multibillion-dollar company (Brinn and Page, 1998).

Search engines are not the only new business spawned by network research. A rapidly evolving industry has developed around social networks. Products are developed based on information gleaned from mapping out people's social links. The businesses range from facilitating business contacts to providing dating services. Some of the companies aim to revolutionize the sales process by identifying the shortest acquaintance path from a salesperson to a target person. In the past 3 years more than 20 companies have emerged that use some aspects of social networks to provide benefits to consumers.

Small companies are springing up to apply an understanding of network structure to practical concerns. One example is the work by Internet Perils, Inc., to improve the robustness of a company's Internet connectivity by identifying bottleneck hubs that lie on seemingly diverse paths. The interest in network research has resulted in a number of network analysis tools. The majority of these tools focus on network visualization; some are free, others can be purchased.

In an intermediate time frame, biology is likely to benefit from advances in network research (Barabási and Oltvai, 2004), and a rush is on to capitalize on applications of the genomics revolution. It is increasingly apparent that the design of successful drugs for complex diseases like cancer or depression depends on mapping out the interactions between cell components. Several companies are involved in commercializing this mapping process, and developing tools that take advantage of network representations of a cell. For example, Genomatica, a San-Diego-based company, has developed a series of tools: Starting with knowledge of the structure of a metabolic network, Genomatica generates predictions useful in a range of ways, from developing drugs to developing strains of bacteria with special metabolic characteristics.

All in all, the committee finds that research on networks not only underlies the affordability and reliability of the glo-

BOX 2-1
Books Relevant to Network Science

General Audience

Albert-László Barabási, *Linked: The New Science of Networks*. Cambridge, Mass.: Perseus Publishing, 2002.

Mark Buchanan, *Nexus: Small Worlds and the Groundbreaking Science of Networks*. New York, N.Y.: W.W. Norton, 2002.

Bernardo A. Huberman, *The Laws of the Web: Patterns in the Ecology of Information*. Cambridge, Mass.: MIT Press, 2001.

Duncan J. Watts, *Six Degrees: The Science of a Connected Age*. New York, N.Y.: W.W. Norton, 2003.

Monograph and Proceedings

Eli-Ben Naim, Hans Frauenfelder, and Zoltan Toroczkai, *Complex Networks (Lecture Notes in Physics)*. Springer-Verlag, October 16, 2004.

S.N. Dorogovtsev and J.F.F. Mendes, *Evolution of Networks: From Biological Nets to the Internet and WWW*. Oxford, England: Oxford University Press, 2003.

S. Bornholdt and H.G. Schuster, eds., *Handbook of Graphs and Networks: From the Genome to the Internet*. Weinheim, Berlin: Wiley-VCH, 2003.

Pedro L. Garrido and Joaquín Marro, eds., *Modeling Complex Systems*, Seventh Granada Lectures, Spain 2002. American Institute of Physics Conference Proceedings, Vol. 661. Melville, N.Y.: AIP, 2003.

Duncan J. Watts, *Small Worlds: The Dynamics of Networks Between Order and Randomness*. Princeton, N.J.: Princeton University Press, 1991.

Graph Theory/Algorithms

Béla Bollobas, *Random Graphs*, 2nd Ed. Cambridge, England: Cambridge University Press, 2001.

Clifford W. Marshall, *Applied Graph Theory*. New York, N.Y.: Wiley-Interscience, 1971.

Joel Spencer, *The Strange Logic of Random Graphs: Algorithms and Combinatorics*. New York, N.Y.: Springer-Verlag, 2001.

Internet/www

Romualdo Pastor-Satorras and Alessandro Vespignani, *Evolution and Structure of the Internet: A Statistical Physics Approach*. Cambridge, England: Cambridge University Press, 2004.

Pierre Baldi, Paolo Frasconi, and Padhraic Smyth, *Modeling the Internet and the Web: Probabilistic Methods and Algorithms*. England: John Wiley & Sons, 2003.

Martin Dodge and Rob Kitchin, *Mapping Cyberspace*. New York, N.Y.: Routledge, 2001.

Martin Dodge and Rob Kitchin, *Atlas of Cyberspace*. England: Addison-Wesley, 2001.

Social Networks

Stanley Wasserman and Katherine Faust, *Social Network Analysis: Methods and Applications*. Cambridge, England: Cambridge University Press, 1994, reprint 1999.

Malcolm Gladwell, *The Tipping Point: How Little Things Can Make a Big Difference*. Boston, Mass.: Little, Brown and Company, 2000.

Per Hage and Frank Harary, *Island Networks: Communication, Kinship and Classification Structures in Oceania*. Cambridge, England: Cambridge University Press, 1996.

Manfred Kochen, *The Small World*. Norwood, N.J.: Ablex Publishing Corporation, 1989.

R.R. McNeill and William H. McNeill, *The Human Web: A Bird's-Eye View of World History*. New York, N.Y.: W.W. Norton, 2003.

Peter R. Monge and Noshir S. Contractor, *Theories of Communication Networks*. New York, N.Y.: Oxford University Press, 2003.

Wayne E. Baker, *Networking Smart: How to Build Relationships for Personal and Organizational Success*. Available online at http://Backinprint.com.

Wayne E. Baker, *Achieving Success Through Social Capital: Tapping Hidden Resources in Your Personal and Business Networks*. Jossey-Bass, 2000.

Economic Systems/Political Networks

Manuel Castells, *The Internet Galaxy*. New York, N.Y.: Oxford University Press, 2001.

Ross Dawson, *Living Networks: Leasing your Company, Customers, and Partners in the Hyper-Connected Economy*. Upper Saddle River, N.J.: Prentice Hall, 2003.

Dirk Messner, *The Network Society: Economic Development and International Competitiveness as Problems of Social Governance*. Frank Cass Publishers, 1997.

Chris Westland, *Financial Dynamics: A System for Valuing Technology Companies*. New York, N.Y.: John Wiley & Sons, 2003.

Networks in the Arts and Culture

Alistair Reynolds, "Glacier" in *The Year's Best Science Fiction 2001*, Gardner Dozois, ed. New York, N.Y.: St. Martin's Griffin, 2002.

Mark Lombardi, Robert Hobbs, and Judith Richards, *Mark Lombardi: Global Networks*. Independent Curators, August 2003.

Mark C. Taylor. *The Moment of Complexity: Emerging Network Culture*. Chicago, Ill.: University of Chicago Press, 2002.

Other Books Discussing Various Aspects of Networks

Fritjof Capra, *The Web of Life: A New Scientific Understanding of Living Systems*. New York, N.Y.: Anchor Books/Randomhouse, 1996.

Geoff Mulgan, *Connexity: How to Live in a Connected World*. Cambridge, Mass.: Harvard Business School Press, 1998.

Steven Strogatz, *Sync: The Emerging Science of Spontaneous Order*. New York, N.Y.: Hyperion, 2003.

Judy Breck, *Connectivity: The Answer to Ending Ignorance and Separation*. Lanham, Md.: Rowman & Littlefield, 2004.

bal communications, transportation, and power infrastructures but also is an important source of economic growth via the creation of new commercial endeavors.

Finding 2-7. Recent network research is leading to new and growing businesses.

In summary, human understanding of networks has the potential to play a vital role in the 21st century, which is witnessing the rise of the Connected Age. There is an enormous demand for information on how to design and operate large global networks in a robust, stable, and secure fashion. In subsequent chapters, the committee discusses the dearth of fundamental scientific knowledge that would ensure that outcome. In Chapter 3, the committee looks at the use of networks in the military, and the special attention given in the statement of task to their role in network-centric warfare.

REFERENCES

Albert, R., and A.L. Barabási. 2002. Statistical mechanics of complex networks. Reviews of Modern Physics 74(1): 47–97.

Amaral, L.A.N., A. Barrat, A.L. Barabási, G. Caldarelli, P. De Los Rios, A. Erzan, B. Kahng, R. Mantegna, J.F.F. Mendes, R. Pastor-Satorras, and A. Vespignani. 2004. Virtual round table on ten leading questions for network research. The European Physical Journal B 38(2): 143–145.

Amin, M. 2002. Modeling and control of complex interactive networks. IEEE Control Systems Magazine 22(1): 22–27.

Arquilla, J., and D. Ronfeldt. 2001. Networks and Netwars. Santa Monica, Calif.: RAND.

Barabási, A.L. 2002. Linked: The New Science of Networks: The Future of Terror, Crime, and Militancy. Cambridge, Mass.: Perseus.

Barabási, A.L., and R. Albert. 1999. Emergence of scaling in random networks. Science 286(5439): 509–512.

Barabási, A.L., and Z.N. Oltvai. 2004. Network biology: Understanding the cell's functional organization. Nature Reviews: Genetics 5(2): 101–114.

Boccara, N. 2004. Modeling Complex Systems. New York, N.Y.: Springer.

Bower, J.M., and H. Bolouri. 2001. Computational Modeling of Genetic and Biochemical Networks. Cambridge, Mass.: MIT Press.

Brinn, S., and L. Page. 1998. The anatomy of a large-scale hypertextual Web search engine. Computer Networks and ISDN Systems 30(1–7): 107–117.

Buchanan, M. 2002. Nexus: Small Worlds and the Groundbreaking Science of Networks. New York, N.Y.: W.W. Norton.

Diamond, J. 1999. Guns, Germs and Steel: The Fates of Human Societies. New York, N.Y.: W.W. Norton.

Dorogovtsev, S.N., and J.F.F. Mendes. 2002. Evolution of networks. Advances in Physics 51(4): 1079–1187.

Dorogovtsev, S.N., and J.F.F. Mendes. 2003. Evolution of Networks: From Biological Nets to the Internet and WWW. Oxford, England: Oxford University Press.

Huberman, B.A. 2001. The Laws of the Web: Patterns in the Ecology of Information. Cambridge, Mass.: MIT Press.

Jasny, B.R., and L.B. Ray. 2003. Life and the art of networks. Science 301(5641): 1863.

Koonin, E.V., Y.I. Wolf, and G.P. Karev. 2002. The structure of the protein universe and genome evolution. Nature 420(6912): 218–223.

Newman, M.E.J. 2003. The structure and function of complex networks. SIAM Review 45(2): 167–256.

Ottino, J.M. 2004. Engineering complex systems. Nature 427(6973): 399.

Perez, C. 2002. Technological Revolutions and Financial Capital: The Dynamics of Bubbles and Golden Ages. Cheltenham, England: Edward Elgar Publishers.

Proceedings of the National Academy of Sciences of the United States (PNAS). 2004. Mapping Knowledge Domains. 101(Suppl.1).

Rheingold, H. 2002. Smart Mobs: The Next Social Revolution. Cambridge, Mass.: MA Basic Books.

Strogatz, S. 2001. Exploring complex networks. Nature 410(6825): 268–276.

Watts, D.J. 2003. Six Degrees: The Science of a Connected Age. New York, N.Y.: W.W. Norton.

Watts, D.J. 2004. The "new" science of networks. Annual Review of Sociology 30(1): 243–270.

Watts, D.J., and S. Strogatz. 1998. Collective dynamics of "small-world" networks. Nature 393(6684): 440–442.

3

Networks and the Military

As the Army, in fact all of the services, the Joint Staff, and the Department of Defense (DOD) look to the future, a new vocabulary dominates the planning as well as the strategic and tactical direction of the entire military process: Doctrine, Organization, Training, Materiel, Leadership/Education, Personnel, and Facilities. "Information dominance and superiority," "net-centricity," "network-centric warfare," and "network-centric operations" are frequently used terms that have become part of the lexicon associated with transformation to a future military force.

NETWORKS AND THE ARMY

From its earliest days, the Army has moved through doctrine, training, and equipping the forces relying on some form of networked communications. For the most part this was an Army Signal Corps function satisfied by switches, radios, satellites, and cable. Army leadership wanted to be sure it could talk to whomever it needed and left decisions about the network to technically competent "communicators."

This paradigm has shifted dramatically. Leaders of all military services and DOD have become aware that a successful doctrine for warfare in the Information Age demands that they engage network issues at many different levels. Force transformation is seen to depend on the development of a coherent system of interacting networks using rapidly evolving enabling technologies.

The ability of joint and coalition units to integrate and maintain "connection" has been essential to operations in Iraq and Afghanistan. Joint Network Nodes (JNN) allow for direct connectivity between warfighters of different services and the Global Information Grid (GIG), thus enabling Army brigade and battalion units to stay connected to the Joint Task Force (JTF) in support of their mission. Blue Force Tracker (BFT), which allows warfighters on the ground to answer the questions Where am I? Where are my buddies? was a great success in Iraq. JNN and BFT were rapidly followed to the Afghanistan and Iraq battlefields by Joint Combat Identification (JCI) systems, improvised explosive device (IED) countermeasure systems, artillery locating systems, persistent surveillance and dissemination systems, mine detection systems, and long-range scout surveillance capabilities (day/night and all-weather), all coming together to provide what the Army calls "battle command on the move."

One only has to examine the key suite of technologies for command, control, communications, computers, intelligence, surveillance, and reconnaissance systems—C4ISR systems—to realize that none of the systems can be effective standing alone. To win the battles of the future, the integration and networking of C4ISR systems is essential, from concept development to combat in the field.

At the same time as these C4ISR technologies and systems provide a manifold improvement in combat capabilities they provide a manifold problem of complexity. None of the systems stands alone on the battlefield. Most C4ISR systems were developed in separate "stovepipe" programs by hard-working, imaginative, and productive engineering teams; yet all must interoperate to varying degrees. Further complicating the issue will be adapting the highly centralized and hierarchical command structures of the Army (and other service forces) and accommodating both old and new generations of technology. Particularly vexing, for example, will be requirements to network unmanned vehicles, including remote sensors and weaponry, while keeping responsible and accountable human beings in the loop.

NETWORK-CENTRIC WARFARE AND NETWORK-CENTRIC OPERATIONS

The leadership of DOD has believed for some time that global communications technology, epitomized by the Internet and the World Wide Web, will fundamentally transform the conduct of war in the 21st century just as airpower transformed it between World Wars I and II. This belief is embedded in two strategic assumptions of profound significance to the Army. First, that better situational awareness and communication in combat situations will result in higher combat effectiveness. This implies facile and high-bandwidth communications between elements of all the services in combat operations as well as shared information in a common format. Second, it is assumed that better situational awareness will make forces more mobile by virtue of allowing heavy armor to be replaced by agility. These assumptions underlie the notion of a transformation of U.S. military forces by bringing them from the Industrial Age into the Information Age. They are captured in the strategic concept of network-centric warfare (NCW), which has four main tenets (Garstka and Alberts, 2004):

- A robustly networked force improves information sharing and collaboration.
- Such sharing and collaboration enhance the quality of information and shared situational awareness.
- This enhancement, in turn, enables further self-synchronization and improves the sustainability and speed of command.
- The combination dramatically increases mission effectiveness.

In DOD today, the network is seen as perhaps the most potent aspect of this change. It captures the essence of the ongoing transformation and is a central element in improving combat effectiveness. According to LTG Steve Boutelle, U.S. Army Chief Information Officer, the Secretary of Defense has said that the single most transforming thing in our force will not be a weapons system, but a set of interconnections (Military Information Technology, 2003). Thus, early in its deliberations the committee developed

Finding 3-1. DOD and all the military services have a vision of the future in which networks play a fundamental role.

Definition and implementation of the concept of NCW is the goal of the DOD Office of Force Transformation (OFT). Information about the initiative may be found on the OFT Web site,[1] where the concept is described as "an emerging theory of war in the Information Age" that "broadly describes the combination of strategies, emerging tactics, techniques, and procedures, and organizations that a networked force can employ to create a decisive war fighting advantage." It is said to have "applicability for the three levels of warfare—strategic, operational, and tactical—and across the full range of military operations from major combat operations to stability and peacekeeping operations." But the devil is in the details. Defining and implementing the concept has proven to be a huge challenge.

As DOD has worked to come to grips with the definition of NCW, the concept has evolved into a more encompassing notion, "network-centric operations" (NCO). The latter is also described on the OFT Web site.[2] NCO is based on revised tenets that are designed so that the hypotheses underlying them can be tested experimentally based on field data acquired in case studies. Relative to the tenets of NCW, the tenets of NCO emphasize the use of shared information by social networks. This evolution is documented in *Network Centric Operations Conceptual Framework Version 2.0* (Garstka and Alberts, 2004) posted on the Web site, which contains the most current definitions of both NCW and NCO.

NCW encompasses three domains of activity: physical, information, and cognitive. NCO adds a fourth, the social domain, and in addition emphasizes policies and procedures in the cognitive and social domains that lead to effective use of the information provided by the physical and information domains (Garstka and Alberts, 2004).

Finding 3-2. DOD has recognized the value of cognitive and social domains in NCO.

"The physical domain is where strike, protect and maneuver take place across the environments of sea, air and space. It is where the physical infrastructure that supports force elements exists. The key elements of the physical domain are (1) the network and (2) net-ready nodes" (Garstka and Alberts, 2004, p. 49). "The information domain is where information is created, manipulated, value-added and shared. It can be considered the 'cyberspace' of military operations. The key elements of the information domain are (1) data and (2) information" (Garstka and Alberts, 2004, p. 49).

"The cognitive domain is where the perceptions, awareness, understanding, decisions, beliefs, and values of the participants are located" (Garstka and Alberts, 2004, p. 23). A key process in this domain is "sensemaking," which requires the participants to construct effective mental models of a situation in which they find themselves. The military has formulated a model of how sensemaking occurs and how it can be influenced by information technology (Gartska and Alberts, 2004, pp. 29–37).

Finally, "the social domain is where people, organizations, practices and cultures intersect" (Garstka and Alberts, 2004, p. 26). *Conceptual Framework Version 2.0* identifies the attributes of networked structures and cultures, of network-centric people, and of how they collaborate, heavily emphasizing the dependence of combat effectiveness on performance in the cognitive and social domains.

Finding 3-3. Current DOD investments in network research include no activity in the cognitive and social dimensions of NCO, specifically in the vital area of decision making in an information-rich environment.

---

[1] At http://www.oft.osd.mil. Accessed September 1, 2005.

[2] At http://www.oft.osd.mil/initiatives/ncw/ncw.cfm. Accessed September 1, 2005.

The value of NCW is said to be greatest at the intersection of the four domains. Analysis of recent military operations in Iraq and Afghanistan suggests, however, that only the information domain is represented (OFT, 2005). Box 3-1 contains observations from an analysis of two of a number of case studies commissioned by OFT to evaluate the value of NCO. What seems clear from battlefield reports and analyses is that the present systems used by the Army and other services need to be improved and integrated into a solution encompassing the physical, cognitive, and social domains,

---

BOX 3-1
Case Studies in Net-centric Operations

The Office of Force Transformation commissioned a series of case studies to evaluate the value of network-centric operations. In one study—U.S./U.K. Coalition Combat Operations during Operation Iraqi Freedom—an attempt was made to judge the value-added for the Force XXI Battle Command Brigade and Below (FCBC2)/Blue Force Tracker (BFT) system used in conjunction with existing C4 capabilities.

The final analysis of the case study noted the following four points:

- FBCB2/BFT improved coalition operations, although in a somewhat limited way, by giving coalition units situational awareness of one another.
- The limited deployment, training, usage, and operation of FBCB2/BFT with the U.K. units constrained its contribution to overall situational awareness.
- The perception that U.S. forces did not use FBCB2/BFT in interfacing with U.K. forces discouraged subsequent use of the system between coalition forces.
- Anecdotally, the greater benefits appeared to be at the operational and strategic levels of command, where blue force feeds from multiple sources were aggregated to provide a coalition common operational picture (COP).

Specific combat actions extracted from the case study provide a clear view of the conclusions of the analysis:[1]

At one objective when the U.S. forces were attempting to secure a bridge on the River Euphrates, 1 Brigade Combat Team (BCT) was to secure the bridgehead to allow 2 BCT to be the breakout force. When forward elements of 1 BCT were reaching the objective the plan was that lead elements of 2 BCT should be four hours behind them. It should be noted that the formations were out of radio contact. In fact, elements of 2 BCT were up to eighteen hours behind according to the time and space calculations made by units of 1 BCT based on the situational awareness afforded by FBCB2/BFT. Hence, the assault on the objective became a hasty defense until such time as the operation could be conducted. This demonstrates the utility of FBCB2/BFT to allow a unit to synchronize its actions with the operational context and conform to the collective scheme of maneuver. Furthermore it demonstrates how the 1 BCT commander was provided time to consider new courses of action.

Interviews with personnel from 1 (UK) Armored Division highlighted that planning that had been undertaken prior to crossing the line of departure regarding the use of FBCB2/BFT had not resulted in the system being used as agreed between unit commanders. . . . Apparently, when the U.S. forces were engaged by Iraqi forces south-west of Baghdad, the system was disregarded and the relief in place conducted through the more familiar use of liaison officers on the ground.

In both of these extracted vignettes we see that the focus is solely on the relay of information. Thus, the network is nothing more than a network of linked communications devices. Granted that provides a commander with increased situational awareness; but, despite that increased situational awareness, the commander does not derive a new approach to warfare. For example, the 1 BCT commander conducted a hasty defense, which is exactly what he would have done when faced with the delay of 2 BCT. While it could be argued that increased use and familiarity with the FBCB2/BFT system would have resulted in greater reliance on the system, it is noteworthy that the forces portrayed in the second vignette resort to direct human contact with liaison officers.

The fourth point of the analysis—that the system's greatest benefit may be at operational and strategic levels—is particularly revealing. It suggests a return to a more traditional perspective, when an overall commander could assume an overview of the battlefield from a nearby hillside. (This implies that the scope of the battle may be more geographically distributed in modern warfare, but there are no new elements to the mission of defeating the opposing force.)

A similar case study was conducted in which U.S. Air Force war games were evaluated to examine the difference in the kill ratios of fighter planes equipped with voice-only communications and those with link-16 data communication capabilities. The study demonstrated that aircraft sharing the greater amount of information—link 16—had kill ratios more than twice those of aircraft that were equipped voice only. The significance of this finding is that the metric used is not something derived from a new theory of warfare; kill ratios are as old as warfare itself.

SOURCE: Adapted from Garstka and Alberts (2004).

---

[1]Garstka and Alberts, 2004, p. 5-5.

as well as the information domain. This is an elusive objective that involves both near- and far-term efforts in network research, and includes efforts that go beyond research per se.

Finding 3-4. Current DOD investments in network science and technology are almost exclusively in the information domain of NCO.

CHALLENGES

Upon assuming his post, the Secretary of the Army focused on issues of immediate concern to the Army: "A network-centric capable force is one that is robustly networked [and] fully interoperable and shares information and collaborates by means of a communications and information infrastructure that is global, secure, real time, reliable, Internet-based, and user-driven."[3] This vision of NCO cannot be delivered with the knowledge or technology available today.

Challenges associated with present-day military information networks can be found at the tactical, operational, and strategic levels (Garstka and Alberts, 2004). They include the following:

- Lack of overall integrating architectures and systems engineering for enterprise networks;
- Inadequately trained, educated, and certified personnel and network users;
- Network management and lack of joint network configuration management;
- Network security and information assurance;
- Requirements to model, simulate, and test large networks before deployment;
- Fusion of multiple sensors and sensor types across the network for real-time decision making;
- Design of individual service networks and synthesis with the DOD GIG;
- Understanding the relationship between network structure and complexity and its impact on organizational design and individual and unit behaviors; and
- Energy-efficient electronics to reduce soldier loads and simplify logistics support.

These challenges cannot be met with current technology alone. While significant resources have been expended to develop network-centric capabilities, most of the improvements are based on existing technology rather than on new results from network research. Even today, given all the resources that have been devoted to creating a networked force, there are no scientifically based guidelines for either the specification of the infrastructure needed or its design.

---

[3]S.W. Boutelle, chief information officer, Department of the Army, "Remarks by the Honorable Francis J. Harvey, Secretary of the Army, December, 2004, as quoted in Slide 5: The Way Ahead," briefing to the committee February 2005.

Bandwidth availability is often cited as the central issue. During Operation Iraqi Freedom, only the strategic level was truly bandwidth-enabled. It is possible that network research would lead not merely to advances in bandwidth utilization but also to efficient applications that would better support the networked force at the operational and tactical (warfighter) levels of combat. DOD is projected to spend $30 billion to $50 billion over a relatively short time frame to transform and develop a network-centric force. It is possible that benefits could be realized faster and at lower cost with appropriate research in key areas.

Planning and experimentation for NCO focuses on relatively narrow scenarios taken from past war experience and fails to reflect the reality of determined 21st century adversaries. Present-day information technologies on which military networks depend are extremely vulnerable to attack and manipulation. More importantly, by limiting the focus of NCO to the physical, information, cognitive, and social domains, technologies based on networks in non-C4ISR domains likely to be of importance in the future are being overlooked.

Finding 3-5. Although no specific biological domain is identified for emphasis in NCO, the Army is responsible for critical defense operations in this domain that involve multiple interacting networks.

A military operation consists of a myriad of activities occurring in sequence and parallel, with each activity associated with one or more networks that bear on the outcome. A biological attack on the United States, perhaps the "mother of all nightmares," is an extreme but excellent example. The multiplicity of networks involved in such a real-world contingency illustrates the Army's central role, as well as the wide applicability of benefits likely to ensue from investments in network research. Box 3-2 describes how Army units, not necessarily girded for combat, would be required to lead the nation in reacting to a biological attack. Such an attack would provide a true test of the effectiveness of net-centric operational principles.

OPTIMIZING WARFIGHTING ORGANIZATIONS

A number of times in history, the deployment of new weapons (from cannons to armored tanks to nuclear bombs) has conferred a decisive military advantage. But at other times, the most important military advantages have come not from new weapons but from new ways of organizing fighting forces. The Greek phalanx, the armored cavalry squadron, and a unit of action such as the modular brigade are all examples of military organizations that, when they were first used, conferred decisive military advantages. Today, the dramatically reduced costs of communication allowed by new information technologies are making it possible to organize military forces in very different ways. This

> **BOX 3-2**
> **Dependence of Army Operations on Networks: An Example**
>
> Whereas the Future Combat System emphasizes physical networks, future Army operations will actually depend on many other types of networks. Figure 3-2-1 shows a highly simplified version of specific networks that might be involved in the response to a biological attack against the United States. The national response quickly diverges into civilian and military chains. On the civilian side, the first task is to identify the causative agent—a procedure involving biological networks. The public health response then relies on various types of social networks to track the disease and control its spread, as well as prepare for a potential follow-on attack.
>
> On the military side, the response is shown across the staff areas used by the military. (The "J"—for Joint—numbers are standard military designators for the respective staff areas of responsibility shown.) The schematic illustrates how in future Army operations all the network types described in Table 2-1—physical, social, and biological—would be involved. For this simplified schematic, only two network types per staff area are shown. Clearly, as with the civilian response discussed above, each task would be supported by many subtasks, most relying on various additional network types.
>
> To illustrate the potential power of networks in combat operations, Figure 3-2-1 draws on some futuristic concepts. For example, under J-3 Operations, note the use of "community networks," derived from biology. This example is drawn from theoretical work currently being pursued by numerous researchers on the adaptation of animal behavior algorithms to support autonomous swarms of unmanned aerial vehicles (UAVs). Using algorithms from flocking/swarming behavior and ant path determination, a flock of UAVs could be launched on a mission. If some of them were destroyed en route, the flock could self-organize and re-form to continue the mission.
>
> In the case of J-5 Strategic Plans and Policy, there is the suggested use of "ecological networks," also derived from biology. In this instance, one could envision the use of information taken from our understanding of ecology to develop plans and policy for warfighting in areas where we have little specific knowledge of the terrain or environment. Knowledge about how networks influence the terrain or environment could improve war planning. Similarly, knowledge about group-forming networks could speed the establishment of coalitions needed for coalition campaigns. It is evident that while not necessarily even realizing it, the military draws on networks and network theories that have been developed in various fields. A fuller understanding of these underlying fields thus holds considerable potential.

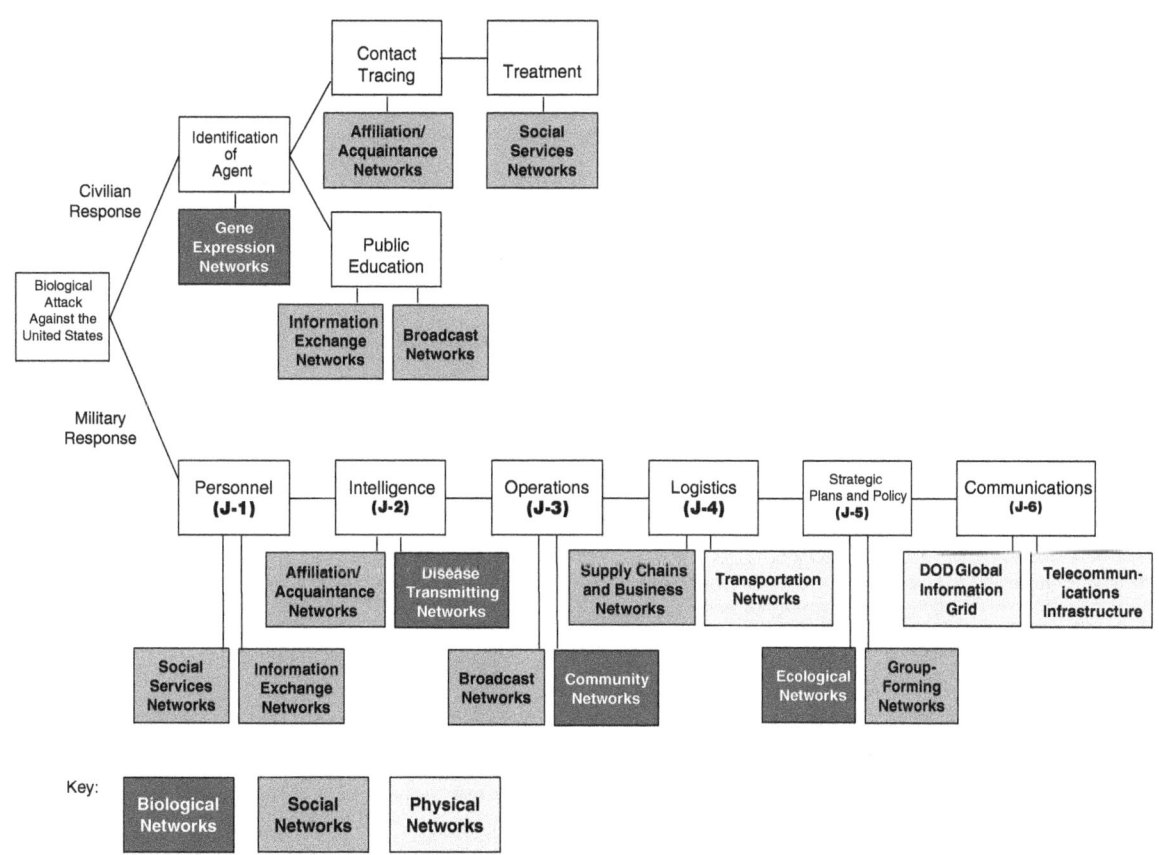

FIGURE 3-2-1 Representative activities and networks involved in responses to a bioterrorist attack.

possibility already has been realized in the concept of "netwar" (Arquilla and Ronfeldt, 2001).

Twenty-first century communications technology makes it possible for every battlefield element to be connected with every other battlefield element, including individual warfighters, sensors and weapons, and vehicles and aircraft, manned and unmanned. Such connectivity has been shown to enable real-time situational awareness and a common operational picture of the battlefield. These and other connections to such things as remote artillery or an aerial weapons platform, greatly extend the capabilities of the individual warfighter.

But increased situational awareness alone is likely to be of limited value if nothing else changes in the military command and control structure. For instance, soldiers who are aware of their situation but unable to make decisions using that information are unlikely to be much more effective than soldiers without such situational awareness. On the other hand, the vastly increased amount of battlefield information that is now potentially shareable makes possible radically new forms of organization such as loose networks of highly autonomous soldiers who swarm over promising targets without any centralized authority telling them to do so (Arquilla and Ronfeldt, 2001).

Finding 3-6. With the increasing importance of terrorist networks, information warfare, and other unconventional means of combat, the decisive advantages in 21st century wars may arise not from superior weapons but from superior ways of organizing warfighters.

Developing such new organizational concepts, however, requires more than incremental improvements to existing military doctrine. It demands substantial creativity and invention, applied in this case not to creating new physical devices but rather to new organizational forms. And this invention is greatly helped by a rigorous understanding of organizational possibilities in other kinds of systems—for example, businesses, social networks, and biological systems (Malone et al., 2003; Malone, 2004; Olson et al., 2001).

NETWORK RESEARCH OF SPECIAL INTEREST TO THE MILITARY

Table 3-1 summarizes major challenges identified by both the Army and the committee during the course of its study. It lists research areas and objectives in all categories of military operations and highlights the broad potential of network research to support C4ISR and other advances and develop-

TABLE 3-1 Network Research Areas

Research Area	Key Objective	Time Frame	Commercial Interest	Priority for Army Investment
Modeling, simulating, testing, and prototyping very large networks	Practical deployment tool sets	Mid term	High	High
Command and control of joint/combined networked forces	Networked properties of connected heterogeneous systems	Mid term	Medium	High
Impact of network structure on organizational behavior	Dynamics of networked organizational behavior	Mid term	Medium	High
Security and information assurance of networks	Properties of networks that enhance survival	Near term	High	High
Relationship of network structure to scalability and reliability	Characteristics of robust or dominant networks	Mid term	Medium	Medium
Managing network complexity	Properties of networks that promote simplicity and connectivity	Near term	High	High
Improving shared situational awareness of networked elements	Self-synchronization of networks	Mid term	Medium	High
Enhanced network-centric mission effectiveness	Individual and organizational training designs	Far term	Medium	Medium
Advanced network-based sensor fusion	Impact of control systems theory	Mid term	High	Medium
Hunter-prey relationships	Algorithms and models for adversary behaviors	Mid term	Low	High
Swarming behavior	Self-organizing UAV/UGV; self-healing	Mid term	Low	Medium
Metabolic and gene expression networks	Soldier performance enhancement	Near term	Medium	Medium

ments of interest to the Army and DOD. The time frame for realization, the likely degree of commercial interest, and the value to the Army (priority) of the challenges reflect the knowledge and estimates of the committee at the time of the study. The time frames are for the basic research activities in network science necessary to produce actionable technology investment options and are not for completion of technological implementations.

Finding 3-7. There are many challenges associated with implementing NCO in the Army that can be identified, classified, and prioritized to create an investment strategy for network science.

The transformation of the Army and other services into an effective network-centric force requires disciplined study and research in practically all areas involving networks. Today, there is no coherent discipline for the study of networks. A well-defined science of networks would provide a much more efficient path to a fully capable network-centric force.

Finding 3-8. To exploit the full potential of networks, a viable science of networks is required.

REFERENCES

Arquilla, J., and D. Ronfeldt. 2001. Networks and Netwars: The Future of Terror, Crime, and Militancy. Santa Monica, Calif.: RAND.

Garstka, J., and D. Alberts. 2004. Network Centric Operations Conceptual Framework Version 2.0. Vienna, Va.: Evidence Based Research, Inc..

Malone, T.W. 2004. Network the Future of Work: How the New Order of Business Will Shape Your Organization, Your Management Style and Your Life. Cambridge, Mass.: Harvard Business School Press.

Malone, T.W., R.J. Laubacher, and M.S. Morton, eds. 2003. Inventing the Organizations of the 21st Century. Cambridge, Mass.: MIT Press.

Military Information Technology Online Edition. 2003. Interview with Major General Steven W. Boutelle. Available at http://www.military-information-technology.com/article.cfm?DocID=33. Accessed June 21, 2005.

Office of Force Transformation (OFT). 2005. The Implementation of Network Centric Warfare. Document 387. Available at http://www.oft.osd.mil/initiatives/ncw/ncw.cfm. Accessed June 21, 2005.

Olson, G.M., T.W. Malone, and J.B. Smith, eds. 2001. Coordination Theory and Collaboration Technology. Mahwah, N.J.: Lawrence Erlbaum Associates.

4

The Definition and Promise of Network Science

In preceding chapters the committee demonstrated the importance of networks to society in general and the Army in particular. It documented that there is interest in research on the properties of networks in any number of civilian and military applications. The committee also established that a pressing national demand exists for "the creation of a new field of investigation called network science to advance knowledge of complex systems and processes that exhibit network behavior," as expressed in the statement of task. In this chapter it addresses the question of how network science should be defined and positioned.

WHAT IS NETWORK SCIENCE?

The first item in the statement of task (Box 1-2) asks how the new field of investigation called network science envisioned by the committee should be defined. The committee's research into this seemingly simple question created a surprisingly complex response.

Logically, the notion of network science is straightforward. It is the organized knowledge of networks based on their study using the scientific method. This notion is immediately valuable in that it distinguishes science from technology. Throughout human history technology often evolved far earlier than the scientific knowledge on which it is based. A classic example is the creation of advanced technologies for the production of metal tools and weapons far in advance of the science of metallurgy.

This is the case today for networks. Although the technologies underlying the design, construction, and operation of the global physical communications, information, and distribution networks described earlier are quite advanced, the underlying scientific knowledge has remained rather rudimentary, according to the experts and literature surveyed by the committee. Developing the metallurgy analogy further, the current state of knowledge about physical communication and information networks is similar to the knowledge of metallurgy for weapons and tools in 16th century Europe (Diamond, 1999). Quite sophisticated steel swords and, ultimately, guns were made by entirely empirical processes for creating and forming the steel, without knowing anything about atomic structure, grain boundaries, or the influence of processing on grain boundaries and dislocations. The development of atomic-level metallurgy in the 20th century enabled a quantum leap in the engineering of lightweight, high-strength materials (e.g., for turbine blades and aircraft). However, the weapons created using the empirical technology of the 16th through 19th centuries enabled Europeans to conquer the world during that period, just as modern communications and information technology can and will transform the battlefields of the 21st century. The components of modern communication and information networks are the result of technologies based on fundamental knowledge emanating from physics, chemistry, and materials science. Their assembly into networks, however, is based largely on empirical knowledge rather than on a deep understanding of the principles of network behaviors gained from an underlying science of networks.

In this report the committee examines the state of fundamental knowledge emanating from research on the science of networks rather than the state of empirical knowledge emanating from research on the technologies that go into the construction of physical networks. This is a profound and fundamental distinction that must be appreciated to understand the report. Its flavor may be illustrated by contrasting the concepts of discovery and invention. In a scientific study pursuing new fundamental knowledge, discoveries are made about how the objects of study behave. For example, in the study of the complex three-dimensional network formed by magnetic atoms in solid atomic lattices, the emergent behavior of phase transitions to various ordered states was discovered experimentally and predicted by sophisticated analyses of network models that describe the interactions between the spins on nearby lattice sites (Binney et al., 1992). This is a discovery in the science of networks that illustrates what sorts of network structures and dynamics are required to predict such behaviors. Contrast this to the invention of the protocol stacks used in the Open Systems Interconnection (OSI) and Transmission Control Protocol/Internet Protocol (TCP/IP) reference models for the Internet (Tanenbaum, 2003). These models were invented to produce reliable connections between computers under some assumed conditions. They work when these conditions are satisfied but are not necessarily suitable for substantially different network structures like those needed for interplanetary communications (Jackson, 2005). They do not address such issues as whether there exist conditions under which the networks on which they run might exhibit emergent behaviors, i.e., behaviors not predictable from the known behaviors of their components. This report is devoted to an assessment of the current state and future prospects of scientific discoveries about networks rather than of the improvement of the technologies of the components of physical networks or the invention of methodologies for integrating these components into networks to solve specific problems.

The fact that network science is logically conceivable does not imply, however, that the concept has been realized in practice. The term "network science" evokes dramatically different images in the minds of workers in different applications domains. The communications engineer envisages the knowledge needed to design a complex communications network like the Internet or the telephone system. The sociologist thinks of networks of influence, like boards of directors or certain social organizations. The business person visualizes the study of informal human networks that enable firms to function, like supply networks and influence networks within large organizations. The physicist thinks of the theory of complex systems, focusing on how order emerges from the apparently random interactions between the nodes through phase transitions or self-organization. The power engineer envisages the knowledge underlying the design and control of the electric power grid. The cell biologist contemplates models of genetic and metabolic networks that enable cell function. And so it goes.

The committee addressed this situation by conducting two inquiries. First, it inquired whether there was a body of knowledge widely thought of as being the content of network science that was taught in universities. The results of this inquiry are reported in Chapter 5 and Appendix C. Second, it asked the practitioners of various applications of networks about their notions of network science. The results of this second inquiry are reported in Chapter 6 and Appendix D. What the committee discovered was that practitioners in each major applications area had their own local nomenclatures to describe network models of the phenomena in which they were interested and their own notions of the content of network science. These notions overlap, but are not identical.

Finding 4-1. Today there is no encompassing science of networks reflected in the practices and perceptions of practitioners of network research and development.

That is, the committee's research validated the implication in the statement of task that such a "new field of investigation" has yet to be codified. The committee's point of view is that the operational definition of network science is what the community of researchers who view themselves as working in this field of investigation actually do. Because a coherent community does not exist across the various applications areas, an opportunity exists for the Army to nucleate such a field by its leadership and funding policies. This opportunity is real rather than virtual because a modest consensus exists among network researchers as to what a core "network science" might encompass. The committee developed the substance of this consensus in Chapters 5 and 6. Network science will evolve into whatever its practitioners create. Those two chapters therefore describe the current state of this rapidly evolving area of investigation.

Finding 4-2. The notion of a science of networks is evolving, and there is limited understanding of its ultimate scope and content.

The communities from which network science is expected to emerge encompass many disciplines and applications areas. Today these communities are characterized by a diversity of nomenclature, models, and opinions about which aspects of the topic are most important. New terms and concepts proliferate. Some are common across many fields —for example, across statistics, economics, sociology, and biology. Others are found in only one or a few subfields— for example, molecular biology, neurology, epidemiology, and ecology, all within biology. Some are given different labels in different fields while meaning essentially the same thing. As documented in Chapter 5, there seems to be a widespread realization that codifying a common nomenclature and body of core knowledge would be useful, but this has not yet occurred. This is why the task statement's concept of a "new field of investigation called network science" is both sensible and timely.

Describing the communities of practice from which the science might emerge does not suffice to provide an operational definition of network science. In addition we must describe its scope and content. Given the rapid evolution of research in this area, this must be done at a high level of abstraction. The details will change significantly over time.

The notion of a network is abstracted from the physical, biological, or social realities that are experimentally observed. As discussed in Chapters 5 and 6, a network is described by its structure (e.g., nodes and links), its dynamics (the temporal attributes of nodes and links), and its behaviors (what the network "does" as a result of the interactions among the nodes and links). Thus, a network is always a representation or model of observable reality, not that reality itself. This creates interesting questions about the uniqueness of a specific network representation of a particular phenomenon—for example, the network model of a metabolic process. It is difficult to establish that a successful network model of a social or biological process is unique in the sense that Maxwell's equations uniquely describe the propagation of electromagnetic waves independent of the details of the associated physical environment.

The statement of task asks how a new field of investigation called network science should be defined. Given the evolving notion of a science of networks, any answer to this question will be ephemeral, and the scope and content of network science will evolve as its practitioners develop it. Proposing a formal definition entails two additional risks. First, since the various network application communities perceive this topic in different ways, some of them are likely to criticize, even reject, any definition offered. Second, in light of this possibility, differences of opinion over the definition may become a rationale for discounting the contents of this report. Nevertheless, in the spirit of providing the Army with a framework for thinking about what network science might become, the committee offers the following tentative definition:

Finding 4-3. Network science consists of the study of network representations of physical, biological, and social phenomena leading to predictive models of these phenomena.

By focusing on the development of models and properties of the underlying representations, this new area of scientific investigation offers the promise of developing tools, techniques, and models that apply to multiple applications areas. It also offers the happy prospect of simplifying and codifying a variety of nomenclatures and lexicons. Thus, one may reasonably expect that creation of a field called network science will not only provide a body of rigorous results that improve the predictability of the engineering design of complex networks, but also speed up basic research in a variety of applications areas. (The defining characteristics of a network and its behaviors are explored further in Chapter 6.)

POSITIONING OF NETWORK SCIENCE

Science tells us how the world operates, and technology gives us practical applications of the resulting insights. These make their way into various sectors of society: into the medical tools, procedures, and remedies of our health-care sector, into the products and services of our economy at large, into the texts and classrooms of our educational centers, into the laws and administration of our government, and into the weapons and communications systems of our military. Paths leading to military strength, health-care excellence, a trained labor force and economic vibrancy all follow the flow from science to technology to institutional forms and applications in their plentiful variety.

These paths all have beginnings and endings, with specific tasks and characteristics at different points along the way. They begin in gestation, reach an inflection point and grow rapidly, mature when their application is readily understood and widespread, then ultimately age and decline. The quarters of a life cycle are gestation, growth, maturity, and decline. They control not only biological life, but the lives of nations and economies as well (Perez, 2002).

We see this in a seed that sprouts, flowers, and dies; in a product that passes through research and development (R&D) into the marketplace and then into every home; and in an Army that pioneers systems that spread throughout the military, then on into society at large, passing from the arcane to ubiquity and, ultimately, to obsolescence. Ultimately, all cycles are surpassed by another cycle—be it of organism, product, or weapons system—that is better adapted to the extant surroundings. In this life cycle context, network science is somewhere near the end of its gestation, poised for takeoff and growth in the decade ahead.

When new paradigms first appear, the scientific communities that pertain have little or no social organization. In the growth phase of their cycle, groups of collaborators and "invisible colleges" characteristic of a more mature science develop around their bodies of knowledge. Network science is ready to complete this first phase but is not yet ready to enter its growth stage. The Army can play a crucial role in facilitating the transition now.

Indeed, truly surprising results might arise from a systematic study of network science. For example, it is widely held that a revised military paradigm is needed to address evolving threats and opportunities associated with terrorism at home and abroad. These threats arise from network behaviors, specifically the adaptation of social networks to the increasing capabilities of communication and information networks (Arquilla and Ronfeldt, 2001; Berkowitz, 2003). This adaptive phenomenon has been observed over the centuries. Typically, engineered networks designed with one set of social behaviors in mind are, over time, exploited by disruptive elements (e.g., criminals and terrorists) for their own purposes. This is a general historical pattern, examples of which include disruption of commercial naval shipping by pirates in the 18th century, train robberies in the 19th century, airplane hijackings in the 20th century, and terrorism and cybercrime in the 21st century, including the destruction of the World Trade Center on September 11, 2001. Large infrastructure networks evolve over time; society becomes more dependent on their proper functioning; disruptive elements learn to exploit them; and society is faced with challenges, never envisaged initially, to the control and robustness of these networks. Society responds by adapting the network to the disruptive elements, but the adaptations generally are not totally satisfactory. This produces a demand for better knowledge of the design and operation of both the infrastructure networks themselves and the social networks that exploit them. This demand cannot be met by existing knowledge, because the circumstances that create it were not anticipated when the networks were designed and built.

Finding 4-4. A gap exists between currently available knowledge about networks and the knowledge required to characterize, design, and operate the complex global physical, information, biological, and social networks on which the well-being of our citizens has come to depend.

Closing this gap is an urgent matter, because society has become dependent on the reliable, robust operation of complex global communication, information, transportation, power, and business networks. The disruption or exploitation of these networks by adversarial social networks of terrorists or criminals is a demonstrated threat, making an investment in network science not only strategically sound but also politically urgent.

Finding 4-5. Advances in network science can address the threats of greatest importance to the nation's security.

In summary, the committee finds that although there is not universal agreement on what network science is today, there is an emerging consensus on what it can become tomorrow. Moreover, there is a pressing demand for the fundamental knowledge that can be expected to emanate from such a science. Thus, network science is positioned as an emerging new field of investigation at the beginning of its growth curve and of compelling national interest and one that the Army has a unique opportunity to nucleate. In Chapters 5 and 6, the committee turns to an exposition of the results of its research on the content, status, and challenges of this emerging field. Then, in Chapter 7, it articulates how the Army can create value by nucleating the new field and supporting its growth.

REFERENCES

Arquilla, J., and D. Ronfeldt. 2001. Networks and Netwars: The Future of Terror, Crime, and Militancy. Santa Monica, Calif.: RAND.

Berkowitz, B. 2003. The New Face of War: How War Will Be Fought in the 21st Century. New York, N.Y.: Free Press.

Binney, J.J., N.J. Dowrick, A.J. Fisher, and M.E.J. Newman. 1992. The Theory of Critical Phenomena. Oxford, England: Clarendon Press.

Diamond, J. 1999. Guns, Germs and Steel: The Fates of Human Societies. New York, N.Y.: W.W. Norton.

Jackson, J. 2005. The Interplantery Internet. IEEE Spectrum Online. Available at http://www.spectrum.ieee.org/WEBONLY/publicfeature/aug05/0805inte.html. Accessed August 22, 2005.

Perez, C. 2002. Technological Revolutions and Financial Capital: The Dynamics of Bubbles and Golden Ages. Cheltenham, England: Edward Elgar Publishers.

Tanenbaum, A.S. 2003. Computer Networks, 4th edition. Upper Saddle River, N.J.: Prentice Hall PTR.

5

The Content of Network Science

In this chapter, the committee determines topics for inclusion within the boundaries of network science and network science research as a "new field of investigation." The chapter describes the activities conducted to help define the scope and content of network science and presents the committee's factual findings.

HOW DO WE KNOW?

To arrive at a definition of network science and identify the topics that it might encompass, the committee undertook two activities. First, a team of committee members expert in the domains of engineered, physical, biological, and social networks reviewed available academic courses to determine their topical contents. The common elements of the courses were extracted and taken to be a provisional de facto specification of the topics contained in the core science spanning these diverse applications areas. The team's proposal was then circulated to selected academicians for suggestions and refinement, and the results of this effort are collected in Appendix C.

Second, the questionnaire described in Appendix D was circulated to over 1,000 experts in applications areas pertinent to network science asking them for their definitions of the term and their notions of appropriate topical content. The results from the questionnaire, described in Chapter 6 and Appendix D, confirmed the results of the first effort and assisted the committee to elaborate a working definition of network science (Finding 4-3).

The results from both efforts further revealed that the term "network science" evoked different mental models in different individuals and communities. What follows is an articulation of the common elements of these mental models.

CONTENT

As discussed in Chapter 4, network science means different things to different people. It does not exist today as a coherent field of investigation. The committee was charged with assessing whether turning it into a new field with this name would be feasible and useful. One test of the utility of doing so would be to examine the extent to which current research on networks exhibits a core content that cuts across the diverse applications areas.

The questionnaire results discussed in Chapter 6 reveal that there are common elements in these diverse applications that can help to create an operational definition of the field pertinent to the committee's statement of task. Specifically, there seems to be widespread agreement that the common core of network science is the study of complex systems whose behavior and responses are determined by exchanges and interactions between subsystems across a well-defined (possibly dynamic) set of pathways. The central point is that the behavior of a network is determined both by the pathways (structure) and by the exchanges and interactions (dynamics). Moreover the structure itself may be (and usually is) dynamic. This is a flexible definition that allows flexible interpretation in the various applications domains. It is elaborated upon and extended in Chapter 6.

A central outcome of the committee's work is the realization that network ideas span an enormous range of disciplines and applications domains. As might be expected, researchers in each domain have their own terminologies and lexicons, so communication among them is not always straightforward. There is a growing notion that these dialects mask an underlying commonality, but the nature of this commonality remains fluid.

Finding 5-1. Network science is an emerging discipline whose boundaries are evolving.

For network science to be regarded as a science, it must encompass core principles that can be taught to students. These core principles, generally embedded in quantitative models, should enable predictions of network behaviors given the structure and dynamics of the network as inputs.

These predictions must be testable experimentally so they can either be verified or proven false. Moreover, the core principles and their associated models and tests will need to be captured in a core curriculum that can be communicated to students. The committee found broad agreement among experts in diverse applications domains on a set of core topics that would need to be mastered to pursue a discipline labeled as "network science."

Finding 5-2. There is broad agreement among experts on topics necessary for inclusion as the core content of network science.

The specific topics included in the core content are described in Appendix C. The central notion is that a network is described by its structure and dynamics, which combine to provide a complete specification of its properties (including functions and behaviors).

The structure of a network is specified by indicating which nodes are linked to which other nodes and whether the links are unidirectional or bidirectional. From this information a number of figures of merit characterizing the structure of the network can be determined. Textbooks and major review articles have been written on this topic (Albert and Barabási, 2002; Dorogovtsev and Mendes, 2003; Newman, 2003; Watts, 2004). The calculation of these figures of merit for various classes of structural models for networks is a staple of courses on networks and an essential core ingredient of network science.

The specification of the dynamics of a network is less straightforward because the dynamics tend to be rather different in the various applications areas. One example is the analysis of phase transitions in physical systems—for example, magnetic atoms in solids. Here the dynamics are specified by the interactions between the spins of the magnetic atoms, which typically vary as a function of the distance between them (Binney et al., 1992). In chemistry and biology, network models are used to describe sequences of chemical reactions. The nodes are typically the reactants and products, with the links being their chemical reactions. The dynamics can be specified by logical models, by rate equations, or by stochastic models of individual reactions (Bower and Bolouri, 2001). In sociology, the nodes are typically people and the links are their interactions. The dynamics are often specified by state models in which the state of one person depends on the states of the other persons with whom she/he interacts as well as on some internal predisposition, often specified statistically (Watts, 2004). Thus, the model dynamics that are introduced in a core course typically depend on the classes of applications that the instructor has in mind.

The essence of network science is making testable predictions about the properties of a network once its structure and dynamics have been specified. A body of knowledge about the standard models and tools for analyzing networks has accumulated over time, as indicated in Appendix C. Because these models and tools constitute knowledge that is often reused in multiple applications areas, they are the remaining elements in the core content of network science.

The core content of network science is basic science, currently consisting of simplified models and of techniques that are appropriate for the analysis of small networks that exhibit low topological complexity in the terminology of Table 2-2. The analysis of network structure is more advanced than that of network dynamics. If adequate structural data are available, structure analysis techniques can be applied to larger and more complex networks using available computer tools. The outputs of model analyses in the core content are insight and qualitative understanding, not engineering design.

The specification of architecture and the design of the physical type networks described in Tables 2-1 and 2-2 are the province of engineering applications domains. The structure, dynamics, and function of the biological and social type networks mentioned in the tables are the subjects of basic research. The application of electromagnetic theory to the design of the power grid affords a useful analogy. Even a graduate physics course in electromagnetism is of little direct use in designing the power grids noted in Tables 2-1 and 2-2. The material in the core content of network science is analogous to that taught in graduate and undergraduate courses in electromagnetism.

Finding 5-3. Research contributing to the core content of network science is basic research (6.1) in the DOD classification scheme.

When the demands on network science imposed by its desired applications are compared with the current state of the knowledge about the science described in Appendix C, a yawning gap appears. The applications require validated theories that allow predicting the properties of global-scale networks under stress conditions. Current knowledge consists of simplified models and tools for analyzing relatively small and simple networks. It seems clear to the committee that substantial development of the core content of network science is required for it to become adequate for its intended applications.

Finding 5-4. Significant investment in the development of the core content of network science is required in order to create adequate knowledge to meet current demands for the characterization, analysis, design, and operation of complex networks.

The networks described in Chapter 2 tend to be both large and complex. They are large if they have many interacting components, typically millions or more for physical networks like the Internet, regional power grids, or transistors on a chip. They are complex if their components exhibit known behaviors, but our knowledge of these behaviors does not suffice to predict the behaviors of the network as a whole (Boccara, 2004). Such complex networks are said to exhibit emergent behavior if the behaviors of their components lead to unanticipated—that is, "emergent"—behavior of the network as a whole in the absence of a centralized controller that creates this behavior by design. As an example, the network of transistors on a computer chip is not normally regarded as exhibiting emergent behavior, whereas an ant colony or the World Wide Web is (Boccara, 2004).

It seems to be widely accepted that investment in basic research will be required to describe the behaviors of social and biological networks. A similar call for investment in basic research might appear counterintuitive for technologically advanced physical networks like the Internet or regional power grids. A few moments of reflection reveals, however, that these physical networks, too, exhibit emergent behaviors. The Internet is robust against expected noises but fragile against unexpected ones, like computer viruses (Doyle et al., 2005). Regional power grids fail infrequently but inevitably, under circumstances not anticipated by grid designers and not adequately dealt with by grid power control systems (IEEE Spectrum, 2004). Contrary to the efforts and hopes of the implementers of advanced technologies, the behaviors of complex physical networks are not yet completely predictable. Moreover, spending to improve the technologies in their components will not remedy this situation. Just like the development of radar awaited the basic science of electromagnetism and that of nuclear weapons awaited the discovery of nuclear fission, the ability to control the complex networks in our lives awaits as yet unforeseen discoveries in the science of networks.

Because committees are notoriously inept at developing curricula or specifying the research content of a science discipline, this committee makes no attempt to do either. It offers the analysis given in Appendix C and discussed above as a test of the proposition that network science be regarded as a coherent area of investigation worthy of investment by the Army. The committee believes that network science fully passes this test.

REFERENCES

Albert, R., and A.L. Barabási. 2002. Statistical mechanics of complex networks. Reviews of Modern Physics 74(1): 47–97.

Binney, J.J., N.J. Dowrick, A.J. Fisher, and M.E.J. Newman. 1992. The Theory of Critical Phenomena. Oxford, England: Clarendon Press.

Boccara, N. 2004. Modeling Complex Systems. New York, N.Y.: Springer.

Bower, J.M., and H. Bolouri. 2001. Computational Modeling of Genetic and Biochemical Networks. Cambridge, Mass.: MIT Press.

Dorogovtsev, S.N., and J.F.F. Mendes. 2003. Evolution of Networks: From Biological Nets to the Internet and WWW. Oxford, England: Oxford University Press.

Doyle, J.C., D. Alderson, L. Lun, S. Low, M. Roughan, S. Schalunov, R. Tanaka, and W. Willinger. 2005. The "Robust yet Fragile" Nature of the Internet. Proceedings of the National Academy of Sciences (PNAS) 102(41): 14497–14502.

IEEE Spectrum. 2004. The Unruly Power Grid. IEEE Spectrum August 2004: 22–27.

Newman, M.E.J. 2003. The structure and function of complex networks. SIAM Review 45(2): 167–256.

Watts, D. 2004. The "new" science of networks. Annual Review of Sociology 30(1): 243–270.

6

Status and Challenges of Network Science

Although no universal consensus exists in the research community that a field of investigation called network science exists today, many researchers describe their work as potentially related to such a field. Many techniques for designing and analyzing networks exist in a variety of application domains, as explored in Chapter 5. In this chapter, the committee draws upon responses to a questionnaire to identify a suite of common characteristics and concerns that span these domains. The questionnaire was circulated to active researchers, identified from literature studies, recursive tracing of collaborative ventures, conference attendance, mailing lists, and interviews with the authors of recent books and reviews. First, the committee describes the questionnaire process and the respondents. Then it summarizes the respondents' assessments of the existence, nature, and challenges of a possible field of network science. Further analysis and discussion are provided in Appendix D.

KEY MESSAGES

From an analysis of over a thousand questionnaire responses, the committee extracted four messages:

- There is no universal consensus among researchers that an identifiable field of network science now exists, in part because there is no accepted definition of what the discipline of network science might be.
- Analysis of the definitions put forth and the research interests of the respondents reveals a suite of input attributes and output properties that could constitute a common core of topics underlying network science.
- Researchers across diverse domains share an implicit understanding that a network is more than topology alone. It also entails connectivity, resource exchange, and locality of action.
- Of seven major challenges identified, the most critical involve characterization of the dynamics and information flow in networked systems; modeling, analysis, and acquisition of experimental data for extremely large networks; and rigorous tools for the design and synthesis of robust, large-scale networks.

QUESTIONNAIRE PROCESS

The questionnaire was developed through an iterative process, starting with an analysis of the statement of task. After a beta-test phase, the final version of the questionnaire was posted on the Web from December 20, 2004, to May 31, 2005. The complete text of the questionnaire is contained in Appendix D. It asked for information about the respondents, their work, and their views on "network science" as a field. It also gave respondents the opportunity to provide as much further information as they wished.

The goal of the ensuing questionnaire invitation process was to reach as large, diverse, and representative a sample of the many relevant research communities as feasible within the study's resources. Overall, the findings that surfaced from the responses and that are presented here are consistent with views held by the committee members. This helped to overcome concerns about the depth and breadth of coverage or other limitations in the questionnaire process. Issues such as reaching beyond the basic snowballing effect, detecting hoax responses, and determining the degree of coverage of the researcher community are discussed further in Appendix D.

THE RESPONDENTS

Finding 6-1. Responses to the questionnaire show a diverse and worldwide network research community with shared concepts and concerns.

Finding 6-2. The results of analyzing the questionnaire responses are consistent across diverse subgroups of respondents.

The responding community is diverse in terms of both geographic distribution and the breadth of interests represented. Responses were received from 29 countries and from 39 states in the United States. Fourteen fields were selected by at least 10 percent of the respondents, and, on average, each respondent selected 3.6 fields. The findings presented in this report do not depend significantly on the field of study or the locale of the respondents but could be limited by the fact that most of them (72 percent) are in the academic community. Further details are given in Appendix D.

DISSENTING VOICES

Finding 6-3. Seventy percent of the responses to the questionnaire accept the idea that network science is a definable field of investigation.

The questionnaire analysis reveals a widespread but not universal consensus among the respondents that a definable field of network science exists. When the reasons for saying there is no such field as "network science" are analyzed, they break down into five kinds of concern: the phrase has no coherent definition; it is broad to the point of vacuity; it is too early to define such a field; it is merely a new name for already existing fields; or, defining such a field is the wrong approach. In addition, respondents also indicated that the field suffered from excessive hype. The distribution of these responses is shown in Figure 6-1.

Of the responses, 70 percent were affirmative to question Q3a: "Is there an identifiable field of network science?" Twenty-three percent of respondents answered no, and 7 percent did not answer. These percentages show little dependence on the backgrounds of the respondents.

The pervasiveness of dissenter concerns across the responding communities reinforces the need for a clear definition of the field of network science, anchored in the expressed approaches of the researchers involved. It also reinforces the idea that care must be taken not to overstate what is achievable in such a field. More positively, articulating an explicit definition of the term "network science" may address some of these concerns.

DEFINING THE FIELD

The first question in considering a possible field of network science is this: What are its contents and scope? Questions 3a and 3b directly address this issue and provide empirical data on the nature of network science as practiced by current researchers. The responses to four other questions proved highly relevant. Further analysis is presented in Appendix D.

The committee structured its analysis in terms of two basic questions: What are the defining attributes of a network? What are the derived properties of interest? If these questions have answers that are common across many application domains, then network science might be identified as the insights, lexicon, measurements, theories, tools, and techniques that allow one to map between desired output properties of a given network and its input attributes. Mapping is needed in both directions: (1) determining the output properties that arise from specific input attributes and (2) determining the input attributes that could be designed into a new network or achieved by intervention in an existing network in order to realize particular output properties.

If network science is to exist in a meaningful way, these approaches also must be effective over many application domains, with well-understood techniques to apply general tools, methods, and models to specific domains. As a hypothetical example, one might envision a simulation tool that deals with network models across a wide range of size scales and timescales, with a growing suite of model libraries customized to specific application domains—for example, ecological networks, metabolic networks, transportation networks, and so on.

Attributes of a Network

Finding 6-4. Analysis of the responses reveals three common attributes of networks: (1) they consist of nodes connected by links, (2) nodes exchange resources across the links, and (3) nodes only interact through direct linkage.

Few responses captured all three attributes, but all three appear consistently, either explicitly or implicitly (in more domain-specific entries), across a wide range of subject domains. The percentage of responses in which an attribute was mentioned explicitly is indicated in Figure 6-2. For brevity, these attributes are designated "connectivity," "exchange," and "locality":

- *Connectivity.* A network has a well-defined connection topology in which each discrete entity ("node" in graph-theoretic terminology) has a finite number of defined connections ("links") to other nodes. In general, these links are dynamic.
- *Exchange.* The connection topology exists in order to exchange one or more classes of resource among nodes. Indeed, a link between two nodes exists if and only if resources of significance to the network domain can be directly exchanged between them.
- *Locality.* The exchanged resource is delivered, and its effects take place, only in local interactions (node to link, link to node). This locality of interaction entails autonomous agents acting on a locally available state.

These attributes are discussed in more detail in Appendix D.

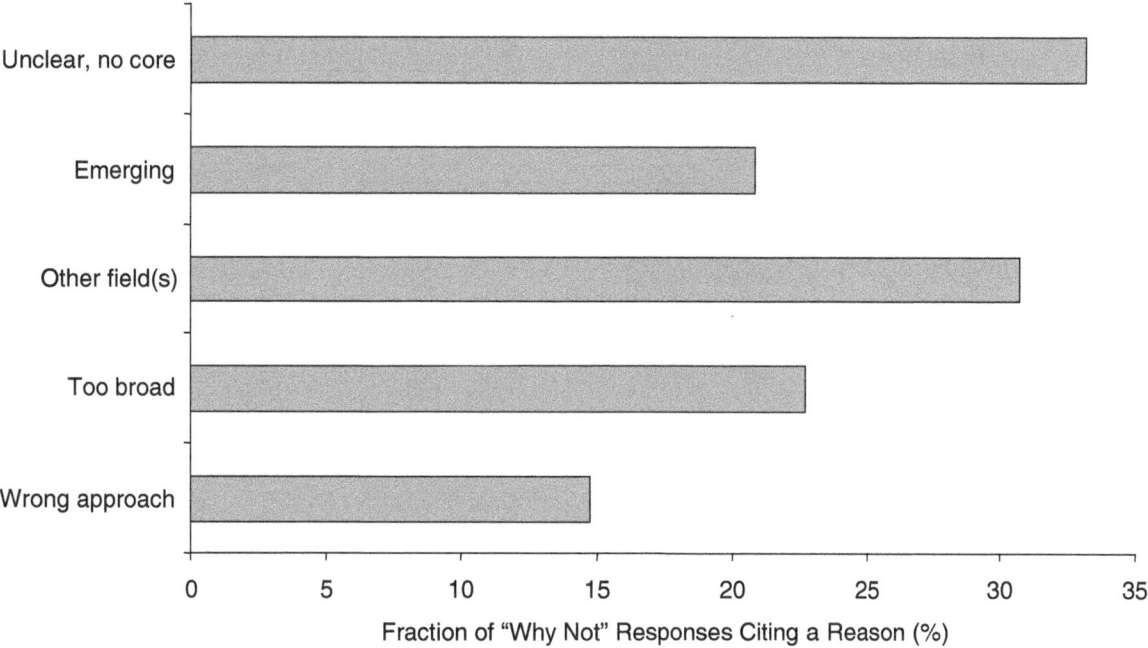

FIGURE 6-1 Reasons for saying there is no field of network science.

Derived Properties of Networks

Finding 6-5. Respondents expressed a need for network science to provide tools that answer a common set of questions across a broad range of applications.

Thirty-three percent of the responses provided definitions relating to the output properties of networks. Analysis of the proposed definitions identified six output properties that spanned a wide range of application domains: characterization, cost, efficiency, evolution, resilience, and scalability. However, only 7 percent of the responses explicitly mentioned classes of tools to address the derivation of these properties. The most frequently mentioned were modeling, simulation, and optimization. These themes also appear in the respondents' research challenges, discussed below.

The responses that proposed driving applications for network science pointed to a highly disparate set of applications, generally tightly bound to five major communities of research: technological, biological, social sciences, interdisciplinary, and physical sciences and mathematics. The distribution of these responses is indicated in Figure 6-3.

The analysis of the questionnaire responses also identified three significant problem dimensions that account for the difficulty of many associated challenges and of the research effort required to address them: complexity, the wide range of interacting scales, and network-to-network interactions.

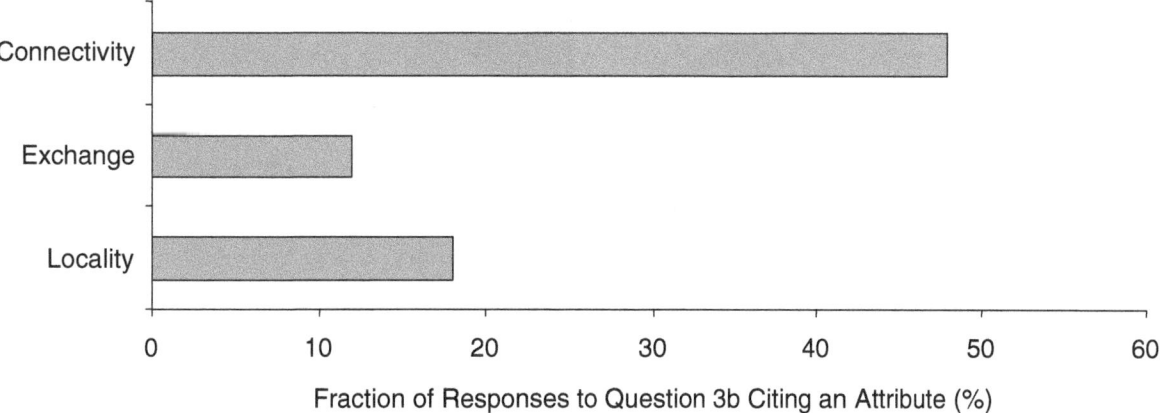

FIGURE 6-2 Share of responses that mention an attribute.

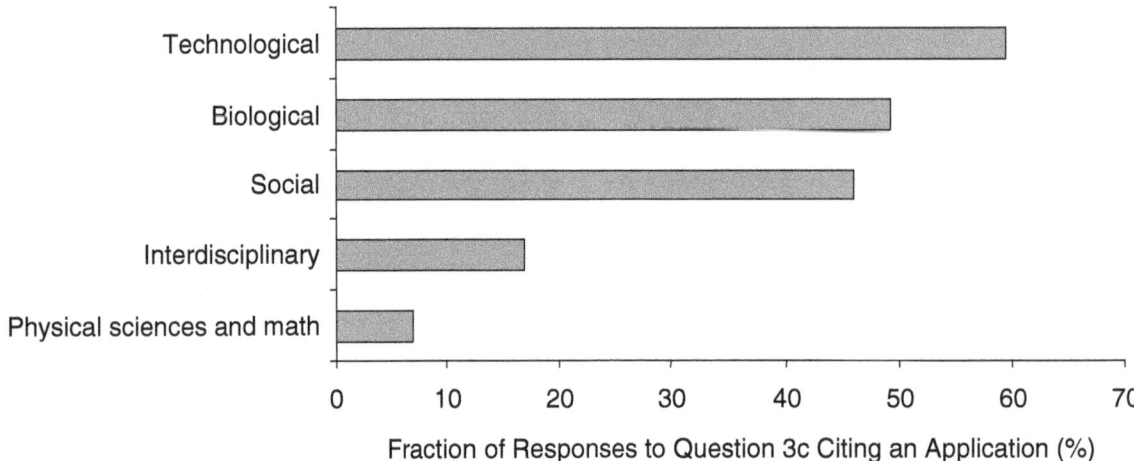

FIGURE 6-3 Responses identifying driving applications.

Future Evolution of the Definition of Network Science

Networks and their associated research programs could be classified and analyzed based on any one of three descriptive categories: input, output, or problem dimension. It is also possible that the categories could become the basis for more precise formal definitions of network science. The questionnaire responses provide evidence that there is a recognizable, coherent common core to the research already being done on network science, and that the scope of the nascent field is both narrow enough to study and deep enough to capture concerns that recur across a diverse range of application domains.

RESEARCH CHALLENGES

Finding 6-6. Respondents identified seven major challenges requiring substantial future work (Figure 6-4).

- *Dynamics, spatial location, and information propagation in networks.* Better understanding of the relation-

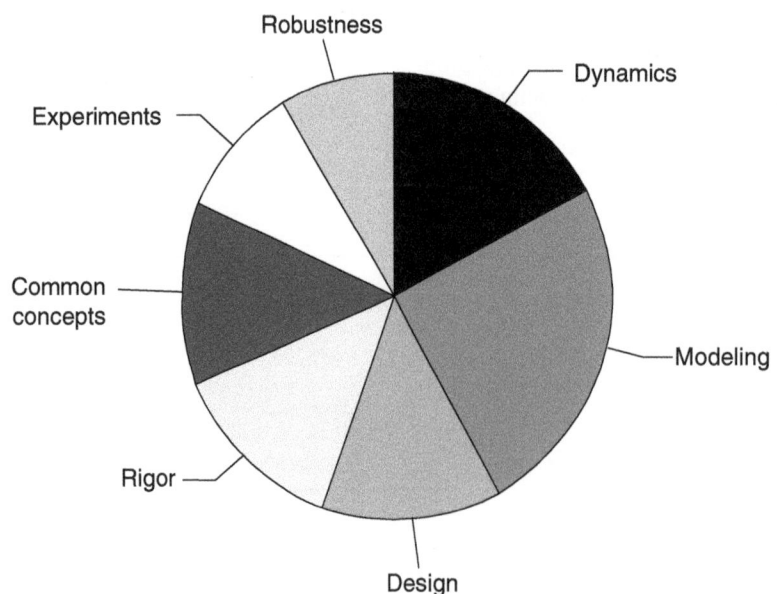

FIGURE 6-4 Major research challenges.

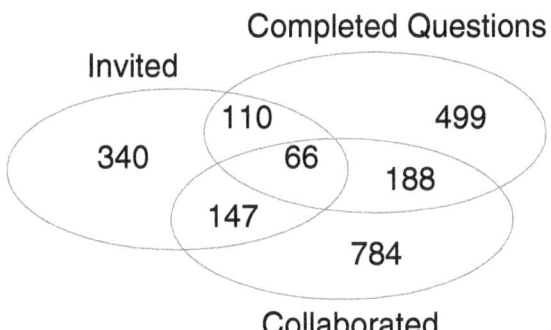

FIGURE 6-5 Relationships among invitees, respondents, and collaborators. SOURCE: Visualization prepared at the committee's request by K. Börner and W. Ke, InfoVisLab at Indiana University.

ship between the architecture of a network and its function is needed.
- *Modeling and analysis of very large networks.* Tools, abstractions, and approximations are needed that allow reasoning about large-scale networks, as well as techniques for modeling networks characterized by noisy and incomplete data.
- *Design and synthesis of networks.* Techniques are needed to design or modify a network to obtain desired properties (such as the output properties discussed in the section "Derived Properties of Networks").
- *Increasing the level of rigor and mathematical structure.* Many of the respondents to the questionnaire felt that the current state of the art in network science did not have an appropriately rigorous mathematical basis.
- *Abstracting common concepts across fields.* The disparate disciplines need common concepts defined across network science.
- *Better experiments and measurements of network structure.* Current data sets on large-scale networks tend to be sparse, and tools for investigating their structure and function are limited.
- *Robustness and security of networks.* Finally, there is a clear need to better understand and design networked systems that are both robust to variations in the components (including localized failures) and secure against hostile intent.

THE SOCIAL STRUCTURE OF NETWORK SCIENCE

The questionnaire data were provided to Katy Börner, associate professor of information science at Indiana University, for analysis of the visible social structure of research

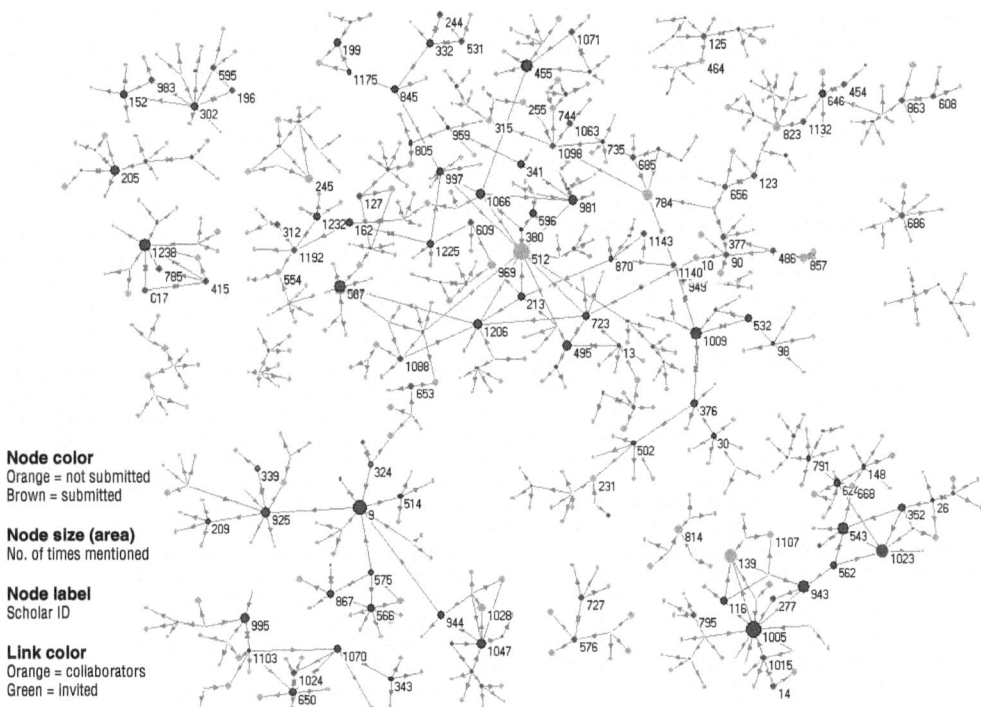

FIGURE 6-6 Network science researchers network. SOURCE: Visualization prepared at the committee's request by K. Börner and W. Ke, InfoVisLab at Indiana University.

in network science. Her analysis considered the 1,241 unique names of network science researchers identified during the course of the questionnaire and citation studies.[1] Names were replaced by unique identification numbers to preserve the anonymity of the respondents. Relationships among the initial invitees, respondents, and identified collaborants are depicted in Figure 6-5.

Figure 6-6 shows the major components (connected graphs of size greater than or equal to 10 nodes) of the resulting network science researcher network (NSRN). The Pajek shows exactly 630 of the 1,241 unique researchers and their association with collaborations and invitations to complete the questionnaire plot (Batagelj and Mrvar, 1997). Each researcher is represented by a node. The nodes are color coded to identify researchers who submitted (brown) or did not submit (orange) questionnaires. The size of the circle (node) reflects the number of times the researcher is mentioned by other researchers.

The details of the visualization analysis are provided in Box D-1 of Appendix D. Upon reviewing the results of the analysis, the committee agreed to include the following two findings on the empirical state of the proposed field of network science:

Finding 6-7. Analysis of the social and collaboration networks of the respondents provides additional evidence that network science is an emerging area of investigation.

Finding 6-8. Analysis of the social and collaboration networks of the respondents provides additional evidence of the multidisciplinary nature of network science.

REFERENCE

Batagelj, V., and A. Mrvar. 1997. Pajek: Program Package for Large Network Analysis. University of Ljubljana, Slovenia. Available at http://vlado.fmf.uni-lj.si/pub/networks/pajek/. Accessed August 18, 2005.

---

[1]Katy Börner, associate professor, Indiana University, "Mapping the expertise and social network of network science researchers," briefing to the committee on April 13, 2005.

7

Creating Value from Network Science: Scope of the Opportunity

In earlier chapters the committee discussed the definition, content, and research challenges of network science. In this chapter it focuses on how investments in network science can create value for the nation in general and for the Army in particular.

CREATING ECONOMIC VALUE FROM RESEARCH KNOWLEDGE

Investments in basic and applied research—that is, in "science"—create new knowledge. They also produce trained research personnel, and they may generate intellectual property (e.g., patents). They do not generate economic value delivered directly to an end customer or user. Rather, a long value chain of activities separates the creation of new knowledge at the beginning of the chain from useful commercial or military applications at the end of the chain (Duke, 2004). Therefore, knowing how the results of research might be used is central to assessing the ultimate value the Army can derive from supporting research in network science.

The concept of real options analysis is helpful here (Amram and Kulatilaka, 1999; Boer, 2002; Mun, 2002). Applying options analysis to research activities allows for research to be viewed as creating the opportunity but not the obligation (here is where the term "option" comes in) to use the knowledge generated to create a final capability that a customer, in this case the Army, is willing to pay a supplier to create and deliver. These options can be valued using standard techniques of real options analysis (Mun, 2002).

Many commercial firms plan their R&D investments using this methodology (Boer, 1999). While the committee does not pursue the full financial analysis discipline in this report, the underlying concepts allow it to assess the scope of opportunity available to the Army as it evaluates its investment alternatives for creating a new science of networks. This assessment is presented below.

SCENARIOS FOR VALUE CREATION

To explore the opportunity for different kinds of options that the Army might decide to create, the committee assumed that the Army will make finite investments in research to advance network-centric warfare (NCW) capabilities and constructed three scenarios that represent fundamentally different levels of investment. After briefly describing these three distinctly different scenarios, the committee presents its findings about how the Army can create value by supporting the development of network science. Since network science does not now exist, the committee had to make some assumptions about its future evolution in order to assess its potential value. All three scenarios involve moving targets, but the nature of these targets differs from one scenario to another. The descriptions that follow provide a sense of the direction dictated by each scenario and the choices available to the Army. Details of the scenarios are provided in Appendix E.

Scenario 1, Building the Base

Scenario 1 involves a modest level of funding (~$10 million per year) that fits into the Army's current scheme for 6.1 basic research. Small amounts of Army risk capital funds are invested to create a knowledge and personnel base from which it can attack the practical problems that arise when trying to provide NCW capabilities. It is for this reason the scenario is called "building the base."

Because the anticipated investment is too small to fund significant interdisciplinary efforts, it should be focused on leveraging existing research in areas related to network science. As discussed in Chapters 5 and 6, the core of network science is rapidly evolving, and Scenario 1 would help it to crystallize.

The committee envisions that the research efforts would be located at major research universities. An important aspect of the program might be a once-a-year conference at an Army laboratory or facility, where the principal investigators (PIs) would report on accomplishments during the year. The program needs enlightened management to support interdisciplinary work accomplished through the interaction of a diversity of PIs. The essence of the program would be the achievement of fundamental advances in network research based on, among other things, statistical physics, applied mathematics, and the development of mathematical models of social phenomena by generously funding only exceptionally talented individuals who are collectively organized into a national network.

Such a program would be the first to address the needs of network science per se. It would be devoted to the study of networks as coherent entities characterized by their architecture, structure, and dynamics. By deliberately adopting a broad theoretical and methodological focus, the program would encourage the creation of fundamentally novel ideas. A wide diversity of approaches can be a key feature of long-term success. Keeping the goals broad and flexible would allow the Army to cultivate such diversity, whereas narrowly defining the program would eliminate much of the creative potential for breakthroughs and new ideas.

The Army's needs are broad and fundamental in nature: It must learn how to approach the creation of a predictive description of large, interacting, layered networks. A basic science program is the first step toward building the critical mass of talent needed to address specific Army problems in this area. This modest approach would allow the Army to identify the relevant research community and organize it so that, in time, it could be called upon to address more specific needs.

The proposed approach differs from existing programs in agencies such as the National Science Foundation (NSF) and the National Institutes of Health (NIH) in that it focuses on network science per se. While a significant amount of research is taking place in communities addressing the applications of networks, almost none of this research is funded by dedicated network science programs.

As a consequence of its discussions with Army and DOD representatives, the committee has come to realize that the fundamental problems underlying effective network-centric operations (NCO) lie in the social domain. Yet how people interact and utilize technology or make decisions based on shared knowledge are areas almost unexplored in the Army's current basic research portfolio. Applications to biology, engineering, and the physical sciences are also essential to Army applications, but the Army is already funding research in these areas. The committee suggests that, on the margin, the most significant problem is not how to build better satellites, tanks, or medicines, but rather how to organize millions of individuals to collect intelligence, deliver supplies, and prosecute wars over an increasingly global and constantly shifting geographical and political playing field (Garstka and Alberts, 2004). This is a monumental problem that has not, however, traditionally been the province of science. Rather it has been managed through a mixture of intuition, experience, and tradition. A significant fraction of the proposed program should address this organizational problem the way scientific problems are addressed: through a combination of theoretical modeling, data analysis, and controlled experimentation.

In Scenario 1 (Appendix E) the committee indicates promising research topics in four broad areas: network structure, network dynamics, network robustness and vulnerability, and network services. Each area has theoretical, empirical, and experimental components. A basic research investment in each of these areas of network science would provide value for the Army. The committee also offers suggestions for improving the return on investment by modest changes in the way that basic research in network science is managed.

Scenario 2, Next-Generation R&D

Scenario 2 envisages applying best practices in industrial R&D management to the Army's investments in projects that combine basic and applied network science. Specifically, the committee expects the objective of these projects to be the articulation of technology investment options that could be exercised by the Army and its vendors to provide a desired capability. The amount of this investment is envisaged to be between $25 million and $100 million annually, roughly $25 million per project. There are expected to be investments in the university community for the basic research and in both Army in-house activities and commercial firms for the applied research. The committee envisages, however, that the R&D projects would be managed in a way profoundly different from the way in which current Army in-house and external centers are managed.

The selection of projects to be funded would be market driven and controlled by a top-level Army team. It is expected that connections between the basic and applied portions of the research will be much more intimate. Modern Internet collaborative tools would be used to manage the day-to-day work in rough analogy to the global design of industrial products. The activities are managed in small, intimate groups devoted to specific subprojects that are integrated into the overall project in a looser networked fashion. People flow from one small group to another over time. The entire team makes up a social network consisting of smaller, more tightly coupled social networks. In short, this scenario envisages the application of modern communications networks and tools and the insights of modern social network theory to transform the management of Army R&D projects.

In Appendix E the committee provides details for market-driven management of such projects. The next-generation R&D model is a new and different approach similar to that of networked organizations like eBay, Intel, and GE. It is based on principles that have worked for many successful companies that needed to get quality products and capabilities to market quickly: Think Big, Start Small, Scale Fast, and Deliver Value. It is this type of next-generation model that can deliver the knowledge, research, and technology that will enable our warfighters to win the nation's wars.

The market-driven approach requires strong commitment from the Army and DOD senior leadership, and from their partners in industry and academia, to make it work. It will be led by a small team of the best and brightest Army "warfighting R&D specialists" committed for a period of 3 or 4 years, highly motivated, and working closely with industry and academia.

The Army has an opportunity for leadership in developing and implementing this new model. At DOD, NCO must be a joint effort, and so should be the new R&D model—after it has been proved in the Army. By moving ahead aggressively to implement this model, the Army can establish itself as the lead for DOD and seize the opportunity to contribute significantly to the improvement of joint network-centric warfighting capability.

The statement of task requests the committee to "identify specific research issues and theoretical, experimental, and practical challenges to advance the field of network science." Briefly stated, three such issues were discussed in Chapters 3–6:

- Current military concepts of "net-centricity" are based on applications of computer and information technology that are far removed from likely results of basic research in network science.
- Current funding policies and priorities are unlikely to provide adequate fundamental knowledge about large complex networks that will advance network-centric operations.
- A basis for a network science is perceived in different ways by the communities concerned with engineered, biological, and social networks at all levels of complexity.

A fourth issue, and major challenge as well, will be to obtain value from the investments that the Army does make to advance network science. In the case of basic research (6.1) alone, the relevant challenges are identified in Scenario 1. In the case of the combined basic and applied research (6.1 6.3) projects envisaged in Scenario 2, the challenges depend sensitively on the topics of the research.

To illustrate the scope and scale of next-generation R&D projects with market-driven management, three projects involving the sociological, engineering, and biological areas of network applications were developed by members of the committee as sample projects for Scenario 2 and are contained in Appendix E. These projects were selected in diverse areas to underline the committee's belief that research will be equally necessary in all areas to advance network science.

The sample project in the social sciences domain outlines a study of local decision making in combat environments. It uses advanced information technology of the sort envisaged for NCO, with the goal of improving the quality of local decisions.

The second project in the engineering domain proposes the design, construction, and testing of a large-area (roughly the size and complexity of a small city) monitoring network for both people and vehicles.

The third describes the construction and testing of a prototype biological surveillance system to detect emerging biological threats. Such a system could also analyze the results of the surveillance and direct appropriate responses.

While all of the sample projects have the potential to advance network science, they should not be construed as a "shopping list," and the committee does not recommend their implementation without careful comparison of their costs and benefits with those of other research projects.

Scenario 3, Creating a Robust Network-centric Warfare/ Operations Capability

The statement of task instructs the committee to "recommend those relevant research areas that the Army should invest in to enable progress toward achieving Network-Centric Warfare capabilities." When the committee examined the literature on this topic, it discovered that the concept of NCW has been superseded in the literature from the DOD Office of Force Transformation (OFT) by an expanded concept, NCO, as described in a conceptual framework document published on the OFT Web site[1] (Cebrowski and Garstka, 1998; Garstka and Alberts, 2004). When members interviewed representatives from the Army and DOD, they found that opinions on NCW and NCO varied widely with regard to both nomenclature and substance. Moreover, the literature on the topic is dynamic, with many new reports and publications. Since this report is intended as an archival document, the committee elected to utilize the published conceptual framework description version 2.0 (Garstka and Alberts, 2004) as point of reference.

Scenario 3 adopts a national point of view. Its purpose is to ask what the nation must do if the strategic vision of NCO is to be implemented. The committee was not tasked to resolve the issues raised in this scenario, but considers their resolution of paramount national urgency.

The committee has stressed that the knowledge of networks that we possess today is not adequate to allow the design of predictable, secure, robust global networks. Members heard presentations and read reports of how the "transformation" to a future force capable of NCO is not likely to be achieved by traditional approaches to creating technology. The committee came to recognize that the policies and

---

[1]For further information, see http://www.oft.osd.mil. Accessed on August 19, 2005.

practices currently used to procure these capabilities do not take into consideration the uncertainties inherent in the current state of understanding the design and implementation of complex networks. The purpose of this brief scenario, then, is to emphasize that the task of designing, testing and operating the envisaged NCO capabilities is of an exceedingly high order of complexity and should be approached as seriously as the Manhattan Project or NASA's race to the moon.

The committee would be remiss in its responsibilities if it failed to note the essential urgency and profound difficulty of this task. The chances of delivering NCO capabilities in a timely and affordable way would be greatly increased by a focused national initiative, combining the initiatives of all services under central leadership, to respond successfully to the diverse challenges of future warfare. Transforming the U.S. military from its current state to that envisaged for NCO as described in the published conceptual framework version 2.0 is the probably the most complex undertaking in the history of the U.S. government (Garstka and Alberts, 2004). It is comparable to the successful pursuit of World War II and the Cold War with the Soviet Union. It is a long-term, difficult, costly, and risky undertaking.

Start by thinking about the task of designing the most complex weapons system built to date—say, a large aircraft carrier. Add to this the complications in the physical domain associated with, for example, secure, reliable wireless communications via satellite to soldiers on a mobile battlefield. In the information domain, add the hardware and software challenges associated with storage, search, and retrieval of orders of magnitude more data in real time, as well as the challenges associated with ensuring the security and reliability of these data. In the cognitive domain, add the issues associated with a junior officer at a local (mobile) workstation processing information from sources at multiple levels in all military services. In the social domain, add the complications of orchestrating the decision-making process in this information-rich, real-time environment and the issues associated with tactics and training to use all this information-processing capability. The committee regards it as highly unlikely that existing methods of designing and procuring weapons systems will be adequate to accomplish this monumental task. Current experience in the services themselves supports this point of view (Brewin, 2005). Further, the committee regards the task of converting the current state of the U.S. military to the vision articulated for NCO as vastly more challenging than seems to be appreciated.

Not only is the task dauntingly complex, the knowledge necessary to accomplish it does not even exist. In similar cases—the Manhattan Project and the initial days of NASA come to mind—a focused, long-term national initiative was required, and it seems likely that something similar will be required in this case also. Thus, Scenario 3 is one in which the United States undertakes a focused national initiative, comparable in scope to the Manhattan Project, to design and deploy NCO capabilities as described in the conceptual framework document 2.0 in all the military services during the coming decade.

Implication of the Scenarios

The main implication of the three scenarios is that there are multiple ways in which the Army can create value by supporting the creation of a science of networks. Which way it selects will depend on circumstances that the committee cannot know.

Finding 7-1. The Army can create value in many different ways from a significant investment in the emerging field of network science.

FINDINGS FROM SCENARIO 1

In Chapter 5 the committee discussed the rudimentary contents of network science. As is often the case, the empirical technology and engineering of large physical networks precede the scientific underpinnings of the technology. This is common throughout history. Humans were making tools and weapons from metals thousands of years before the science of metallurgy was developed. The situation is subtly different for biological and social networks, where the science is devoted to comprehending how these networks function. Tinkering with their natural engineering lies mostly in the future. The "technology" is well developed, but by nature rather than man. In all three cases—physical, biological, and social networks—the technology far outpaces the scientific understanding of what the technology hath wrought.

Finding 7-2. Because network science is at an early stage of its development, a broad portfolio of basic and applied research is expected to create greater value than a more focused portfolio.

Finding 7-3. If there is only a limited amount of funding (e.g., $10 million per year or less), a broad portfolio of basic research is the most promising approach to creating value for the Army.

The main values created by a basic research investment include access to thought leaders (PIs) in the university community, training students through their work on university projects, the development of a community that the Army can access to address its practical problems, and efficient use of research dollars to impact multiple areas of application.

Pursuing research in new ways also can generate value. In order to tackle complex problems, coordination is required. Yet individual insights gained by creative people are usually at the root of the solution of such problems. How does the Army get both at the same time? Network research suggests that small coherent groups associated with exceptional talents can be collected into loosely coupled networks that interact productively without sacrificing the creative potential of individual contributors (Malone, 2004; Watts, 2003). Creating, testing, and refining such an approach to network science by the Army would have far-reaching consequences in other domains.

The Army's network research portfolio must differ from the portfolios of NSF, NIH, and DOE (Department of Energy). This can be done if the Army focuses on the network per se, rather than on specific applications. Perhaps more important, the Army can explore new network approaches to complex problems along the lines indicated above. This notion is distinctly different from that of the centers of research currently pursued by the Army and NSF, among others. An essential element of the network approach, not normally present in centers, is the coherent, coordinated actions of a diverse group of different domain experts to address a precisely formulated, complex problem. This is not just the support of interdisciplinary and collaborative research. It further requires intensely focused attention on a pre-specified problem by diverse groups of domain experts working in concert. Suitable reference models include product development in large firms or the Manhattan Project rather than the efforts that take place at a typical Army lab, NSF center, or DOE user facility.

The conceptual framework for NCO consists of interacting networks in four distinct technology domains: physical, information, cognitive, and social (Garstka and Alberts, 2004). The Army is currently investing primarily, if not exclusively, in R&D associated with a network communications infrastructure and a limited portfolio of applications built upon a network communications infrastructure.[2,3,4] In other words, the current army R&D portfolio spans only two (the physical and the information domains) of the four domains essential to the implementation of NCO.

Whether investments in physical and information infrastructure improve fighting effectiveness, however, depends on what warfighters do with the information available from the infrastructure. Their decisions and actions lie in the cognitive and social domains, which remain unexplored in the current DOD R&D portfolio. The good news is that existing knowledge in these domains could benefit applications in their respective domains of NCO. The bad news is that knowledge in these areas is rudimentary and generic. The insights are qualitative in nature and often not useful for making precise predictions (Malone, 2004; Watts, 2004).[5] Serious investment in both basic and applied research is required before the associated models and concepts can be applied in a predictive way for the development of NCO capabilities for the Army.

Finding 7-4. Since the shift to network-centric operations raises many social and behavioral issues, the Army's network science portfolio should stress developing basic knowledge that could enable applications of network thinking to address the social and cognitive domains.

Because the state of network science is so primitive, the central problems of the field are neither well recognized nor precisely posed. At this stage of its evolution, network science is basic research in the most profound sense: The fundamental questions are still being framed (Watts, 2003). Previous experience in other disciplines (e.g., Einstein's contributions to relativity and quantum theory) suggests that this is a playing field best suited for talent of the highest order, not individuals doing next-step research. For the Army to create value from investments in this area, it must recruit and retain exceptional talent, a difficult task.

Finding 7-5. The Army must find a way to attract the best researchers in network science. This will require stability of funding, the opportunity to interact with a diversity of interesting colleagues, and flexibility to follow the funded research wherever it leads.

Finding 7-6. To attract the best researchers in network science, the Army should fund them to do work that also has applications in nonmilitary areas.

Finding 7-7. To attract the best researchers in network science, the Army must avoid putting restrictions on publications and on foreign nationals.

The committee is well aware that these three findings may appear to be as uncontentious as "motherhood and apple pie." Sadly, this is not the case. In today's global economy, outstanding technical talent has extensive international opportunities. Many of the world's most talented people no longer wish to come to the United States. Many talented people here do not wish to work for the U.S. military. Plentiful opportunities exist for both elsewhere. To create value from basic network research, the Army must attract top talent. The committee believes that for the Army to have a good chance of success in this endeavor, it must heed the three findings.

---

[2]S.W. Boutelle, chief information officer, Department of the Army, "The way ahead," briefing to the committee on February 1, 2005.

[3]J. Garstka, assistant director, concepts and operations, OSD Office of Force Transformation, "Fighting in the networked force: Insights from network centric operations case studies," briefing to the committee on April 14, 2005.

[4]J. Gowens and A. Swami, Army Research Lab, "Army research in network science," briefing to the committee on February 1, 2005.

[5]C.F. Sabel, professor of law and social science, Columbia Law School, "Theory of a real-time revolution," briefing to the 19th EGOS Colloquium, Copenhagen, July 2003.

FINDINGS FROM SCENARIOS 2 AND 3

Valuing technology and managing risks to extract economic value became hot topics in the business literature during the past decade (Boer, 1999; Branscomb and Auerswald, 2001; Chesbrough, 2003; Cooper et al., 2001). It is widely recognized in industry that product development is an inherently non-linear process, full of feedback loops and surprises (Branscomb and Auerswald, 2001; Reinertsen, 1997). Indeed, these same notions apply to DOD basic research (NRC, 2005). Thus, it is plausible to apply the notions developed in the business context to the management of research in network science by the Army.

Finding 7-8. The Army can learn about R&D best management practices from the business sector.

R&D managers in commercial organizations couple their basic research activities with known or anticipated applications. This provides focus and enables a much more rapid time to market (Branscomb and Auerswald, 2001; Chesbrough, 2003; Cooper et al., 2001). The committee believes that the Army could benefit from studying such practices and adapting analogous ones.

Finding 7-9. Additional value can be extracted from the Army's 6.1 basic research investments in network science by coupling them to downstream applied research and to technology development efforts.

Finding 7-10. The results of basic research in network science are more likely to be rapidly put to use to meet Army challenges if the Army also devotes significant resources to related applied research.

These findings are the basis for Scenario 2 in Appendix E, where the committee describes a management process and three sample research projects in the social, engineering, and biological spheres. All are based on and illustrative of current models and tools used for R&D management in the business sector, including software projects (Poppendieck and Poppendieck, 2003). The material presented in Scenario 2 in Appendix E describes how the Army might usefully experiment with new R&D management practices and new ways of integrating its 6.1, 6.2, and 6.3 R&D programs. The three sample projects developed in the scenario also serve to illustrate the truths of Findings 7-9 and 7-10.

Although they can improve the performance of the front end R&D of the Army's network design processes, commercial R&D best management practices must be supplemented by more fundamental change if DOD is to acquire NCO capabilities. R&D is only one step in a complete value chain.

The military procurement value chain is more complex than the chains in commercial firms. Cases in which large commercial firms' best practices for R&D management are known to work well are those in which (1) the science is mature and (2) the market requirements can be determined with fair accuracy. Neither is true in the case of applying the outputs of network science to the procurement of an NCO capability for the Army. Commercial best practices are most likely to succeed when they are applied to the sourcing of information infrastructure. Even here, however, prospects for success are not certain.

There is no "science" at this time that can predict the performance of wireless and wireline communications infrastructures integrated into the architecture for the Global Information Grid (GIG). Science can predict the performance of individual components (e.g., of radios or computers) but not that of the overall system of networks. At best, the applications that warfighters are likely to develop for such a system are almost certain to be surprises conceived and tested in the field before they are embedded into tactics and doctrine.

The current state of sociological models is too rudimentary for them to be applied reliably to simulate uses of such a network a priori. Even the use of simulation to determine the "market" requirements for the physical network is risky, because it lies beyond the scope of current knowledge.

DOD faces a major challenge as it tries to determine how to design a set of interlocking networks of the complexity and scope envisaged for NCO. The committee captured its concerns about the total end-to-end sourcing process in the following finding:

Finding 7-11. The design, testing, and deployment of the overlapping and interacting physical, information, cognitive, and social networks envisioned for network-centric operations concepts are currently beyond the Army's capability. They require a concerted national effort to be achieved in a timely and affordable fashion.

This finding motivated a third scenario, for creating a "robust network-centric warfare/operations capability." It is a response to the committee's hypothesis that the design and procurement of large, complex networks, such as those envisaged for implementing NCO, cannot be done in an affordable fashion using the current practices.

The United States has faced similar challenges in the past—for example, the design of nuclear weapons in the 1940s (LANL, 1986). Responding effectively to such challenges required a focused, coherent, and sustained national effort involving government, industry, and academic partners and the investment of hundreds of millions of dollars annually over a decade or more.

Moreover, recent insights on network organizations suggest that such a national effort must be organized and managed rather differently than a large engineering project (Malone, 2004). Scenario 3 is explored in Appendix E, but only in broad outline, because the committee was not constituted to provide expert advice on this topic. Current knowledge on the organizational implications of network research leads the committee to suggest that the scenario illustrates the direction in which the military should head to obtain the biggest and most certain payoff from investments in network science (Malone, 2004).[6]

REFERENCES

Amram, M., and N. Kulatilaka. 1999. Real Options: Managing Strategic Investment in an Uncertain World. Boston, Mass.: Harvard Business School Press.

Boer, F.P. 1999. The Valuation of Technology: Business and Financial Issues in R&D. Hoboken, N.J.: Wiley.

Boer, F.P. 2002. The Real Options Solution: Finding Total Value in a High-Risk World. Hoboken, N.J.: Wiley.

Branscomb, L.M., and P.E. Auerswald. 2001. Taking Technical Risks: How Innovators, Managers, and Investors Manage Risk in High-Tech Innovations. Cambridge, Mass.: MIT Press.

Brewin, R. 2005. DoD Mulls Network Coordination. Available at http://www.fcw.com/article88939-05-23-05-Print/. Accessed May 31, 2005.

Cebrowski, A., and J. Garstka. 1998. Network centric warfare. Proceedings of the United States Naval Institute 24: 28–35.

Chesbrough, H. 2003. Open Innovation: The New Imperative for Creating and Profiting from Technology. Boston, Mass.: Harvard Business School Press.

Cooper, R.G., S.J. Edgett, and E.J. Kleinschmidt. 2001. Portfolio Management for New Products, 2nd edition. Reading, Mass.: Perseus Books.

Duke, C.B. 2004. Creating economic value from research knowledge. Industrial Physicist 10(4): 29–31.

Garstka, J., and D. Alberts. 2004. Network Centric Operations: Conceptual Framework Version 2.0. Vienna, Va.: Evidence-Based Research, Inc.

Los Alamos National Laboratory (LANL). 1986. Los Alamos 1943–1945: The Beginning of an Era. LASL-79-78 Reprint.

Malone, T.W. 2004. Network the Future of Work: How the New Order of Business Will Shape Your Organization, Your Management Style and Your Life. Cambridge, Mass.: Harvard Business School Press.

Mun, J. 2002. Real Options Analysis: Tools and Techniques for Valuing Strategic Investments and Decisions. Hoboken, N.J.: Wiley.

National Research Council (NRC). 2005. Assessment of Department of Defense Basic Research. Washington, D.C.: The National Academies Press.

Poppendieck, M., and T. Poppendieck. 2003. Lean Software Development: An Agile Toolkit. Boston, Mass.: Addison Wesley.

Reinertsen, D.G. 1997. Managing the Design Factory: A Product Developer's Toolkit. New York, N.Y.: Free Press.

Watts, D.J. 2003. Six Degrees: The Science of a Connected Age. New York, N.Y.: W.W. Norton.

Watts, D.J. 2004. The "new" science of networks. Annual Review of Sociology 30(1): 243–270.

---

[6]C.F. Sabel, professor of law and social science, Columbia Law School, "Theory of a real-time revolution," briefing to the 19th EGOS Colloquium, Copenhagen, July 2003.

8

Conclusions and Recommendations

In this chapter the committee combines its findings into conclusions and offers recommendations. First, it collects the factual findings presented in Chapters 2-7 into three overarching conclusions concerning the importance of networks and the current state of knowledge about them. Next, it articulates specific conclusions that are directly responsive to Items 1 through 3 of the statement of task. Finally, in response to Item 4, the committee provides its recommendations, including for research initiatives. Box 8-1 summarizes how the report responds to the statement of task.

OVERARCHING CONCLUSIONS

Conclusion 1. Networks are pervasive in all aspects of life: biological, physical, and social. They are indispensable to the workings of a global economy and to the defense of the United States against both conventional military threats and the threat of terrorism.

Conclusion 1 was developed in Chapters 2 and 3 and summarized in Tables 2-1, 2-2, and 3-1 and the discussions surrounding them. It sets the stage for the committee's inquiry into the state of knowledge about these networks.

Conclusion 2. Fundamental knowledge about the prediction of the properties of complex networks is primitive.

Given the pervasiveness and vital importance of networks, one might assume that a lot is known about them. As documented in Chapters 5 and 6, however, this is not the case. Although the technology for constructing and operating engineered physical networks is sophisticated, critical questions about their robustness, stability, scaling, and performance cannot be answered with confidence without extensive simulation and testing. For large global networks, even simulations are often inadequate. The design and operation of network components (such things as computers, routers, or radios) are based on fundamental knowledge gleaned from physics, chemistry, and materials science. However, there is no comparable fundamental knowledge that allows the a priori prediction of the properties of complex assemblies of these components into networks. Indeed, such networks are expected to exhibit emergent behaviors—that is, behaviors that cannot be predicted or anticipated from the known behaviors of their components. In the case of social and biological networks, even the properties of the components are poorly known. A huge gap exists between the demand for knowledge about the networks on which our lives depend and the availability of that knowledge.

The committee learned that developing predictive models of the behavior of large, complex networks is difficult. There are relatively few rigorous results to describe the scaling of their behaviors with increasing size. Surprisingly, this is true for common engineered networks like the Internet as well as for social and biological networks.

Simulation rather than analysis is the research tool of choice. In the case of social networks, even simulation is vastly complicated by the diversity and complexity of the agents that are the nodes of the networks—humans or groups of humans "in the wild." Which of their many properties are relevant for developing mathematical models of a particular phenomenon? Existing models of social networks, moreover, represent highly simplified situations and not necessarily ones that are relevant to the Army or network-centric warfare.

Finally, the notion of using network models in biology is relatively new. Controversy swirls around their utility, indeed around that of systems biology itself. In spite of a burgeoning literature on the structure of simple networks, the advancement of the field to allow relating basic scientific results to applications of societal and military interest still lies mostly in the future.

Conclusion 3. Current funding policies and priorities are unlikely to provide adequate fundamental knowledge about large complex networks.

CONCLUSIONS AND RECOMMENDATIONS

BOX 8-1
Summary of Responses to the Statement of Task

The Assistant Secretary of the Army (Acquisition, Logistics, and Technology) has requested the National Research Council (NRC) Board on Army Science and Technology (BAST) conduct a study to define the field of Network Science. The NRC will:

1. Determine whether initiation of a new field of investigation called Network Science would be appropriate to advance knowledge of complex systems and processes that exhibit network behaviors. If yes, how should it be defined?

A working definition of network science is the study of network representations of physical, biological, and social phenomena leading to predictive models of these phenomena. Initiation of a field of network science would be appropriate to provide a body of rigorous results that would improve the predictability of the engineering design of complex networks and also speed up basic research in a variety of applications areas (Chapter 4).

2. Identify the fields that should comprise Network Science. What are the key research challenges necessary to enable progress in Network Science?

General consensus exists among practitioners of network research in diverse application areas on topics that constitute network science (Chapter 5). There are seven major research challenges (Chapter 6).

3. Identify specific research issues and the theoretical, experimental, and practical challenges to advance the field of Network Science. Consider such things as facilities and equipment that might be needed. Determine investment priority, time frame for realization, and degree of commercial interest.

Current military concepts of "net-centricity" are based on applications of computer and information technology that are far removed from likely results of basic research in network science. Table 8-1 lists current areas of network research of interest to the Army, including priority, time frames, and commercial interest (Chapter 3).
Current funding policies and priorities are unlikely to provide adequate fundamental knowledge about large complex networks that will advance network-centric operations. Besides the information domain, there are social, cognitive, and physical technology domains in the current conceptual framework for network-centric operations; there is no "biological" domain (Chapters 2–4).
A basis for network science is perceived in different ways by the communities concerned with engineered, biological, and social networks at all levels of complexity. Basic research efforts are totally incoherent (Chapters 5 and 6).
Options for obtaining value from investments in network science include scenarios ranging from building a base of basic research, to leveraging business practices for market-driven R&D in specific areas of network applications, to creating a robust capability for network-centric operations (Chapter 7).

4. Given limited resources (and likely investments of others), recommend those relevant research areas that the Army should invest in to enable progress toward achieving Network-Centric Warfare capabilities.

Recommendations 1, 1a through 1d, 2, and 3 provide the Army with an actionable menu of alternatives that span the opportunities accessible to it. By selecting and implementing appropriate items from this menu, the Army can develop a robust network science to enable the desired progress (Chapter 8).

NOTE: The statement of task is in lightface; the summary of responses is in boldface.

Fundamental knowledge is created and stockpiled in disciplinary environments, mostly at universities, and then used as required by (vertically integrated) industries to provide the products and services required by customers, including the military. This fundamental knowledge is different in kind from empirical knowledge gleaned during the development of technology and products. You get what you measure. Suppliers of fundamental knowledge measure publications, presentations, students supervised, awards received, and other metrics associated with individual investigators. The knowledge accumulates along traditional disciplinary lines because this is where the rewards are found. Large team activities are relatively rare (except in medicine and large-scale physics experiments) and are mostly left to the consumers of the fundamental knowledge, who must supplement the fundamental knowledge generously with empirical knowledge to convert it into the goods and services desired by the paying customer.

This scheme worked marvelously for more than a half a century, when the United States dominated the world and industries were vertically integrated. With the onset of the global economy in the 1990s, however, the situation began to change dramatically, for a number of reasons. First, knowledge, investment capital, technology, and technical labor are becoming globally available commodities. Second, economic activity, including R&D, is becoming global in scale. Third, these two trends are making the networks on which we depend ever larger and more complex and their susceptibility to disruption ever greater.

This traditional scheme does not work well for generating knowledge about global networks, because focused, coordinated efforts are needed. Thus, there is a huge difference between the social and financial arrangements needed to gain fundamental knowledge about large, complex networks in a global environment and the arrangements that worked so well to provide such knowledge for the design and production of smaller, less complex entities in a national environment. Any successful effort to create the knowledge necessary to secure robust, reliable scalable global networks must come to grips with this reality.

Overall, the committee is led to a view of networks as pervasive in and vital to modern society, yet understood only as well as the solar system was understood in Ptolemy's time. The military has made networks the centerpiece of its transformation effort without a methodology to design networks in the physical and information domains in a predictive way for network-centric operations (NCO). Further, according to the DOD Office of Force Transformation, research in the cognitive and social domains has yet to yield advances comparable to the technological developments in the information domain. At the same time, current efforts by academia to describe networks are fragmented and disjointed. Relatively little of the current research on networks promises to create a science of networks that will generate knowledge adequate to meet the demand.

In short, there is a massive disconnect between the importance of networks in modern society and military affairs on the one hand and, on the other, the support of coherent R&D activities that would raise current network technologies and capabilities to the next level. The Army alone cannot transform this situation, but it can make a beginning.

SPECIFIC CONCLUSIONS

Items 1 and 2 in the statement of task inquire into the appropriateness of a field of investigation called network science and its definition, content, and the research challenges that would characterize it. Elements of a field of network science have begun to emerge in different disciplines spanning engineering, biological, and social networks. The emerging field is concerned with the development and analysis of network representations to create predictive models of observed physical, biological, and social phenomena.

The remarkable diversity and pervasiveness of network ideas renders the study of network science a highly leveraged topic for both civilian and military investment. The provisional consensus around its core content clearly defines the notion of network science. By making an investment in network science, the Army could forge a single approach to a diverse collection of applications.

Conclusion 4. Network science is an emerging field of investigation whose support will address important societal problems, including the Army's pursuit of network-centric operations capabilities.

Although the boundaries of network science are fuzzy, there is broad agreement on key topics that should be included within the field, the types of tools that must be developed, and the research challenges that should be investigated. These were documented in Chapters 3 and 4.

Conclusion 5. There is a consensus among the practitioners of research on networks for physical, biological, social, and information applications on the topics that make up network science.

Responses to its questionnaire greatly assisted the committee in determining "the key research challenges to enable progress in network science." These responses establish that there is a fair degree of consensus on these challenges across practitioners in diverse applications areas.

Conclusion 6. There are seven major research challenges the surmounting of which will enable progress in network science:

- **Dynamics, spatial location, and information propagation in networks. Better understanding of the relationship between the architecture of a network and its function is needed.**
- **Modeling and analysis of very large networks. Tools, abstractions, and approximations are needed that allow reasoning about large-scale networks, as well as techniques for modeling networks characterized by noisy and incomplete data.**
- **Design and synthesis of networks. Techniques are needed to design or modify a network to obtain desired properties.**
- **Increasing the level of rigor and mathematical structure. Many of the respondents to the questionnaire felt that the current state of the art in network**

science did not have an appropriately rigorous mathematical basis.
- **Abstracting common concepts across fields.** The disparate disciplines need common concepts defined across network science.
- **Better experiments and measurements of network structure.** Current data sets on large-scale networks tend to be sparse, and tools for investigating their structure and function are limited.
- **Robustness and security of networks.** Finally, there is a clear need to better understand and design networked systems that are both robust to variations in the components (including localized failures) and secure against hostile intent.

These challenges are elaborated in terms of specific research issues and their theoretical, experimental, and practical difficulties in Chapter 7 and Appendix E within the framework of exploring various investment scenarios. The scenarios respond to Item 3 in the statement of task.

Although all the military services have a vision of the future in which engineered communications networks play a fundamental role, there is no methodology for ensuring that these networks are scalable, reliable, robust, and secure. Of particular importance is the ability to design networks whose behaviors are predictable in their intended domains of applications. This also is true in the commercial sphere. Creation of such a methodology is an especially pressing task because global commercial networks can also be exploited by criminal and terrorist social networks.

Conclusion 7. The high value attached to the efficient and failure-free operation of global engineered networks makes their design, scaling, and operation a national priority.

RECOMMENDATIONS

The statement of task requests investment recommendations from the committee. Options for these recommendations are explored in Chapter 7 and Appendix E. The committee documents in Chapters 2 and 3 that the impact of networks on society transcends their impact on military applications, although both are vital aspects of the total picture. Chapters 3 and 4 explain that the current state of knowledge about networks does not support the design and operation of complex global networks for current military, political, and economic applications. Advances in network science are essential to developing adequate knowledge for these applications.

Recommendation 1. The federal government should initiate a focused program of research and development to close the gap between currently available knowledge about networks and the knowledge required to characterize and sustain the complex global networks on which the well-being of the United States has come to depend.

This recommendation is buttressed by centuries of evidence that disruptive social networks (e.g., terrorists, criminals) learn to exploit evolving infrastructure networks (e.g., communications or transportation) in ways that the creators of these networks did not anticipate. The global war on terrorism, which is a main driver of military transformation, is only one recent manifestation of this general pattern. Society has the same need in other areas, such as control of criminal activities perpetrated using the global airline and information infrastructures. Addressing problems resulting from the interaction of social and engineered networks is an example of a compelling national issue that transcends the transformation of the military and that is largely untouched by current research on networks.

Within this broad context, Recommendations 1a, 1b, and 1c provide the Army with three options:

Recommendation 1a. The Army, in coordination with other federal agencies, should underwrite a broad network research initiative that includes substantial resources for both military and nonmilitary applications that would address military, economic, criminal, and terrorist threats.

The Army can lead the country in creating a base of network knowledge that can support applications for both the Army and the country at large. Maximum impact could be obtained by a coordinated effort across a variety of federal agencies, including the DOD and the Department of Homeland Security, to create a focused national program of network research that would develop applications to support not only NCO but also countermeasures against international terrorist and criminal threats.

Alternatively, if the Army is restricted to working just with the DOD, it should initiate a focused program to create an achievable vision of NCO capabilities across all the services.

Recommendation 1b. If the Army wants to exploit fully applications in the information domain for military operations in a reasonable time frame and at an affordable cost, it should champion the initiation of a high-priority, focused DOD effort to create a realizable vision of the associated capabilities and to lay out a trajectory for its realization.

Finally, if the Army elects to apply the insight from the committee primarily to its own operations, it can still provide leadership in network science research.

TABLE 8-1 Network Research Areas

Research Area	Key Objective	Time Frame	Commercial Interest	Priority for Army Investment
Modeling, simulating, testing, and prototyping very large networks	Practical deployment tool sets	Mid term	High	High
Command and control of joint/combined networked forces	Networked properties of connected heterogeneous systems	Mid term	Medium	High
Impact of network structure on organizational behavior	Dynamics of networked organizational behavior	Mid term	Medium	High
Security and information assurance of networks	Properties of networks that enhance survival	Near term	High	High
Relationship of network structure to scalability and reliability	Characteristics of robust or dominant networks	Mid term	Medium	Medium
Managing network complexity	Properties of networks that promote simplicity and connectivity	Near term	High	High
Improving shared situational awareness of networked elements	Self-synchronization of networks	Mid term	Medium	High
Enhanced network-centric mission effectiveness	Individual and organizational training designs	Far term	Medium	Medium
Advanced network-based sensor fusion	Impact of control systems theory	Mid term	High	Medium
Hunter-prey relationships	Algorithms and models for adversary behaviors	Mid term	Low	High
Swarming behavior	Self-organizing UAV/UGV; self-healing	Mid term	Low	Medium
Metabolic and gene expression networks	Soldier performance enhancement	Near term	Medium	Medium

Recommendation 1c. The Army should support an aggressive program of both basic and applied research to improve its NCO capabilities.

Specific areas of research of interest to the Army are shown in Table 8-1. This table expresses the committee's assessment of the relative priorities for these areas, the time frames in which one might reasonably expect them to be consummated as actionable technology investment options, and the degree of commercial interest in exploiting promising options. Specific research problems and sample projects are given in Appendix E. The committee notes that both trained personnel and promising research problems exist in many of these areas, so the Army should be able to create a productive program readily.

By selecting from among Recommendations 1a through 1c an option that is ambitious yet achievable, the Army can lead the country in creating a base of knowledge emanating from network science that is adequate to support applications on which both the Army and the country at large depend. Regardless of which option (or options) are adopted, Army initiatives in network science should be grounded in basic research.

Recommendation 1d. The initiatives recommended in 1, 1a, 1b, and 1c should include not only theoretical studies but also the experimental testing of new ideas in settings sufficiently realistic to verify or disprove their use for intended applications.

Recommendations 1, 1a, 1b, and 1c span only part of the investment opportunity space—namely, those segments of the space described in Scenarios 2 and 3 in Chapter 7 and Appendix E. They will involve substantial changes in how the Army invests its R&D dollars and in how it plans and manages these investments.

The Army also has the opportunity associated with Scenario 1 in Chapter 7, which involves funding a small program of basic research in network science. This investment of relatively small amounts of Army risk capital funds would create a base of knowledge and personnel from which the Army could launch an attack on practical problems that arise as it tries to provide NCO capabilities.

Investments in basic (6.1) research in network science can generate significant value; however, the committee wants to be crystal clear that such investments have no immediate prospects of impacting the design, testing, evaluation, or sourcing of NCO capabilities. They would create additional knowledge that builds the core content of network science, and they would train researchers who could also be recruited by the Army for later efforts. While the knowledge generated would probably be less valuable than in the case of Scenarios 2 and 3, the cost is less and implementation can be immediate.

If the Army elects to exploit Scenario 1, the committee offers the following two further recommendations:

Recommendation 2. The Army should make a modest investment of at least $10 million per year to support a diverse portfolio of basic (6.1) network research that promises high leverage for the dollars invested and is clearly different from existing investments by other federal agencies like the National Science Foundation (NSF), the Department of Energy (DOE), and the National Institutes of Health (NIH).

This modest level of investment is compatible with the Army's current R&D portfolio. There is an adequate supply of promising research topics and talented researchers to make this investment productive. Additionally, it can be implemented within the Army's current R&D management work processes, although some enhancements along the lines noted in Chapter 7 and Appendix E would improve the return on this investment.

To identify the topics in basic network science research that would bring the most value to NCO, the committee recalls that the open system architectures for computer networks consist of layers, each of which performs a special function regarded as a "service" by the layers above. It is useful to distinguish among the lower (physical and transport) layers of this architecture, the higher (applications) layers that are built on top of them to offer services to the people, and the cognitive and social networks that are built higher still, on top of the services-to-humans layers.

Research on the lower layers of the network architecture is relatively mature. Improving the services offered at these levels is more of an engineering problem than one requiring basic research. The most immediate payoffs from network science are likely to result from research associated with the upper levels of the network architecture and the social networks that are built at an even higher level upon their outputs. This is where the committee thinks that Army investments are most likely to create the greatest value.

An area of particular promise that has little or no current investment is the social implications of NCO for the organizational structure and command and control. Basic research could provide valuable insight into how military personnel use advanced information exchange capabilities to improve combat effectiveness. For example, one might study how troops in combat could use these capabilities to make better decisions. Additional basic research in the core content of network science might help to determine how the Army can most productively utilize the capabilities of its advanced information infrastructure.

Recommendation 3. The Army should fund a basic research program to explore the interaction between information networks and the social networks that utilize them.

The Army can implement Recommendations 2 and 3 within the confines of its present policies and procedures. They require neither substantial replanning nor the orchestration of joint Army/university/industry research projects. They create significant value and are actionable immediately.

The committee's Recommendations 1, 1a through 1d, 2, and 3 give the Army an actionable menu of options that span the opportunity space available. By selecting and implementing appropriate items from this menu, the Army can develop a robust network science to "enable progress toward achieving Network-Centric Warfare capabilities," as requested in the statement of task.

Appendixes

Appendices

A

Biographical Sketches of Committee Members

Charles B. Duke (NAS/NAE), *Chair*, is vice president and senior research fellow in the Innovation Group of Xerox Corporation. He was formerly deputy director and chief scientist at the Pacific Northwest National laboratory. He was founding editor in chief of the *Journal of Materials Research*, the official journal of the Materials Research Society. For a decade he served as editor in chief of *Surface Science* and *Surface Science Letters*. He has served on the governing boards of the American Institute of Physics, the American Physical Society, the Materials Research Society, and the American Vacuum Society. He has written a monograph and over 350 papers on surface science, materials research, semiconductor physics, the electronic structure of molecular solids, and research management. He has edited three volumes on surface science and digital systems integration. He received his Ph.D. in physics from Princeton University following a B.S. summa cum laude with distinction in mathematics from Duke University. He is chair based on demonstrated leadership of research in electronic structures of physical and biological systems and for his expertise in the network design of distributed systems and in real-world applications of control and communication theory.

John E. Hopcroft (NAE), *Vice Chair*, is IBM Professor of Engineering and Applied Mathematics at Cornell University. He received his B.S. from Seattle University and his M.S. and Ph.D. from Stanford University, all in electrical engineering. Dr. Hopcroft is a recipient of the A.M. Turing Award of the Association of Computing Machinery and serves on the National Research Council's Board on Mathematical Sciences and their Applications. His research centers on the theoretical aspects of computing, especially the analysis of algorithms, automata theory, and graph algorithms. He has expertise in computer science and engineering.

Adam P. Arkin is assistant professor of bioengineering and chemistry at the University of California at Berkeley and faculty scientist in physical biosciences at the E.O. Lawrence Berkeley National Laboratory. He is also an assistant investigator at the Howard Hughes Medical Institute. His laboratory develops and applies mathematical theory and computational and experimental approaches to the analysis of cellular function. He was recognized as one of the most innovative young scientists by the MIT *Technology Review* and is a member of the National Academies' Committee on the Frontiers at the Interface Between Computing and Biology. Dr. Arkin received his B.A. from Carleton College and his Ph.D. from the Massachusetts Institute of Technology. He has expertise in quantitative biology and the modeling of cellular networks.

Robert E. Armstrong is a senior research professor in the Center for Technology and National Security Policy at the National Defense University. Dr. Armstrong also serves as a member of the advisory council for the Agronomy Department at Purdue University. He is a colonel in the U.S. Army Reserve. He received his B.A. from Wabash College, his M.A. in experimental psychology from Oxford University, and an M.S. (biology) and a Ph.D. (plant breeding and genetics) from Purdue University. He has expertise in biodefense applications for networks.

Albert L. Barabási is director of the Center for Complex Networks Research at the University of Notre Dame. He is also Emil T. Hofman Professor of Physics at Notre Dame and has previous research experience at the IBM T.J. Watson Research Center. Dr. Barabási has published extensively on complex systems and networks in books, magazines, and journals. He is a member of the Hungarian Academy of Sciences. Dr. Barabási received his M.S. from Eotvos Lorand University in Budapest and his Ph.D. in physics from Boston University. He is author of the book *Linked: The New Science of Networks* and has published the article "Network biology: Understanding the cell's functional organization." His expertise is in biological and physical networks.

Ronald J. Brachman is the director of the Information Processing Technology Office at the Defense Advanced Research Projects Agency (DARPA). He has previous experience at AT&T Labs, where he served as the vice president of Communication Service Research as well as director of the Software and Systems Research Center. Dr. Brachman currently serves as the president of the American Association for Artificial Intelligence and is a fellow of the Association for Computing Machinery. He received his B.S.E. in electrical engineering from Princeton University and his S.M. and Ph.D. in applied mathematics from Harvard University. He has expertise in network applications research.

Norval L. Broome is a consultant for the MITRE Corporation. He was director of San Diego Operations and department head of Naval Communications Systems for MITRE and is a retired engineering duty officer in the U.S. Navy. Dr. Broome's research interests include software-derived multimode, multifunction radios for high-data-rate tactical communications. He received his B.S.E.E. and M.S.E.E. from Purdue University and the Ph.D.E.E. from the California Institute of Technology. He has published in journals of the Institute of Electrical and Electronics Engineers (IEEE), including those of the Antennas and Propagation Society and the Communications Society. His expertise lies in command, control, communications, computers, and intelligence (C4I) systems and network applications.

Stan Davis is writer and consultant on business strategy, organization, and management. He spent 20 years on university faculties, mainly at the Harvard Business School, and was a senior research fellow at the Cap Gemini Ernst & Young Center for Business Innovation. Dr. Davis is author of 13 books, including *Future Perfect*, *Blur*, and *It's Alive! The Convergence of Biology, Information and Business*. He has a Ph.D. in social science from Washington University in St. Louis and has expertise in and knowledge of social and business network applications.

Richard A. De Millo is the John P. Imlay, Jr., Dean of Computing in the College of Computing at the Georgia Institute of Technology. Before accepting his position at Georgia Tech he was the chief technology officer for Hewlett-Packard. Dr. DeMillo's distinguished technology career spans business, government, and academia, including important positions at the National Science Foundation (NSF), Telcordia Technologies, and Purdue University. He received a Ph.D. in information and computer science from Georgia Tech and a B.S. in mathematics from the College of St. Thomas. His expertise is in computer science and engineering.

William J. Hilsman is a retired Army lieutenant general and a consultant to the Institute for Defense Analyses. He is the chief executive officer of DTI International. Before that, he was the CEO and chairman of the board of InterDigital Communications Corporation, which pioneered the development of digital wireless networks. As director of the Defense Communications Agency, General Hilsman managed the National Communications System for the DOD worldwide C3 system. He also was commanding general of the Army Communications–Electronics Research and Development Command. He is a graduate of the U.S. Military Academy and received his M.S. in electrical engineering from Northeastern University. He has expertise in military applications and communications.

Will E. Leland is chief scientist in the Network Systems Research Laboratory of the Telcordia Corporation. As research scientist for Bellcore, Dr. Leland uncovered fundamental characteristics of network analysis and design based on the mathematical concept of self-similarity. He has published extensively on distributed network performance and is the recipient of the Baker and Bennett prizes from the IEEE. He has served as manager of and technical adviser for the Telcordia Network Management System program for the Army's Future Combat Systems and is currently involved with Telcordia's planning and research for the Air Force's Transformational Communications System. Dr. Leland received a B.S. in both mathematics and physics from the Massachusetts Institute of Technology (MIT) and M.S. and Ph.D. degrees in computer sciences from the University of Wisconsin. He has expertise in communications networks and knowledge of research in commercial telecommunications networking and military ad hoc wireless networking.

Thomas W. Malone is the Patrick J. McGovern Professor of Management at the MIT Sloan School of Management. He is also the founder and director of the MIT Center for Coordination Science. Dr. Malone has published four books, most recently *Network the Future of Work*, and over 75 research papers; he also holds 11 patents. Prior to his work at MIT, he was a research scientist at the Palo Alto Research Center of Xerox. He received a B.A. in mathematical sciences from Rice University, an M.S. in engineering-economic systems from Stanford University, and an M.A. and Ph.D. in psychology from Stanford University. His expertise is in human networks, coordination theory, and the use of information technology in organizations.

Richard M. Murray is the chair of the Engineering and Applied Science Division at the California Institute of Technology. His current research is in the area of dynamics and control of mechanical, fluid, materials, and information systems, with applications in aerospace vehicles, robotics, turbomachinery, and thin-film processing. He received a B.S. in electrical engineering from the California Institute of Technology and an M.S. and a Ph.D. in electrical engineering and computer science, both from the University of California at Berkeley. He is an IEEE fellow and a member of the Air Force Scientific Advisory Board. He has published work in several journals, including the paper "Control in an information rich world: Report of the Panel on Future Directions in Control, Dynamics, and Systems." He has expertise in engineering and network controls.

Jack Pellicci is the group vice president of business development for Oracle Corporation's Public Sector Group. He retired from the U.S. Army as a brigadier general with 30 years of experience as an infantry officer, including service as the commanding general of the Personnel Information Systems Command, deputy director of training for the U.S. Army, and brigade commander in the 7th Infantry Division. He is a graduate of the U.S. Army Command and General Staff College, the Army War College, and the Senior Executive Seminar in National and International Affairs. General Pellicci received a B.S. degree in engineering from the U.S. Military Academy at West Point and an M.S.M.E. from the Georgia Institute of Technology. His expertise is in military operations and information system development.

Pamela A. Silver is professor of systems biology at Harvard Medical School and a member of the Department of Cancer Biology at the Dana-Farber Cancer Institute. She has developed novel genetic and cell biological approaches to study the movement of macromolecules in eukaryotic cells, first as a faculty member at Princeton University and now at Harvard. Dr. Silver received a Ph.D. in molecular biology from the University of California and was a postdoctoral fellow at Harvard University. She is an expert in the emerging field of synthetic biology, with a published paper, "The potential for synthetic biology."

Paul K. Van Riper is a retired lieutenant general in the U.S. Marine Corps. He continues to participate in defense and security-related seminars and lectures frequently at the National Defense University and military educational institutions. General Van Riper consults part-time for DARPA and has participated in numerous war games and experiments related to network-centric warfare. He has also served on the Army Science Board and the National Research Council's Naval Studies Board. He received a B.A. from California State College, in Pennsylvania, and is a graduate of the Marine Corps Amphibious Warfare School, the U.S. Army Airborne and Ranger Schools, the U.S. Navy College of Command and Staff, and the U.S. Army War College. He has expertise in military applications for networks.

Duncan J. Watts is an associate professor in the Department of Sociology at Columbia University and was postdoctoral fellow at the Santa Fe Institute and the Sloan School of Management at the Massachusetts Institute of Technology. His research and teaching focus on mathematical and computational modeling of complex networks as applied to problems in social network theory, contagion, computation, and the theory of the organization. Dr. Watts is the author of *Small Worlds: The Dynamics of Networks Between Order and Randomness*; *Six Degrees: The Science of a Connected Age*; and more than 20 peer-reviewed articles, including "The 'new' science of networks." He received a B.S. in physics from the University of New South Wales, Australia, and a Ph.D. in theoretical and applied mechanics from Cornell University. Dr. Watts has expertise in sociological networks.

B

Committee Meetings and Other Activities

MEETINGS

First Committee Meeting, November 15–16, 2004, Washington, D.C.

BAST Network Science Overview
John Parmentola, director, Research and Laboratory Management, Office of the Deputy Assistant Secretary of the Army (Research and Technology)

Scope and Dimensions of Network Science
Charles Duke, vice president, Xerox Innovation Group, Wilson Center for Research and Technology

Dimensions of Network Science
Ronald Brachman, director, Information Processing Technology Office, Defense Advanced Research Projects Agency
Joshua Epstein, senior fellow, The Brookings Institution

Coordination Science as Part of Network Science
Thomas Malone, director and professor, Center for Coordination Science, Massachusetts Institute of Technology

Network Science and Network-centric Operations: An Industry Perspective
Jack Pellicci, group vice president, Oracle Public Sector

Networks Are Ubiquitous
John Hopcroft, Jett IBM professor, Cornell University

Network Science First Thoughts
Richard DeMillo, John P. Imlay, Jr., dean of computing, Georgia Institute of Technology

Collective Dynamics of Small World Networks
Steven Strogatz, professor, Cornell University

Dimensions of Net Science: What Makes a Net?
Will Leland, chief scientist, Telcordia Technologies

Second Committee Meeting, February 1–2, 2005, Washington, D.C.

Army Networks for Net-Centric Operations
LTG Steve Boutelle, chief information officer, Department of the Army

National and International Views on Systems Biology
Adam Arkin, assistant professor, Lawrence Berkeley National Laboratory

Network Science in the New Program in Systems Biology at Harvard University
Pamela Silver, professor, Harvard University

Army Research in Network Science
Anathram Swami, ARL fellow, Army Research Laboratory

Applications of Network Science in Business and Military Organizations
Thomas Malone, director, Center for Coordination Sciences, Massachusetts Institute of Technology

The "New" Science of Networks
Duncan Watts, assistant professor, Columbia University

Third Committee Meeting, April 13–15, 2005, Washington, D.C.

Mapping the Expertise and Social Network of Network Science Researchers
Katy Börner, associate professor, Indiana University

Fighting the Networked Force: Insights from Network-centric Operations Case Studies
John Garstka, assistant director, Office of the Secretary of Defense, Office of Force Transformation

High-Performance Computing for Army R&D in the Bio Sciences and Synergies with C2 Networks
Amado Cordova, RAND Arroyo Center

Network Science and Net-centric Warfare
LtGen Paul van Riper, U.S. Marine Corps (retired)

Fourth Meeting, May 12–13, 2005, Woods Hole, Massachusetts

Writing meeting

TELECONFERENCES

First Full-Message Draft Teleconference, March 31, 2005

Research Initiatives in Network Science, April 13, 2005

Participants
John Doyle, California Institute of Technology
Stuart Milner, University of Maryland

Research Initiatives in Network Science, April 28, 2005

Participant
P.R. Kumar, University of Illinois

C

Content of Network Science Courses

Appendix C offers a detailed account of the results of the survey the committee conducted on the core materials taught in network science courses. To achieve its goal, the committee followed a two-step procedure. First, it searched for currently taught courses on networks, looking in all possible departments, from computer science and physics to the social and biological sciences. A representative list of such courses is provided in Table C-1. The committee's work was facilitated by the fact that many courses post a detailed syllabus on the Web, as well as links to other courses on the same topic. If a syllabus was not available on the Web site, a copy was requested from the instructor. After inspecting the collected syllabi, it was possible to discern a set of core concepts that are shared by a wide range of courses and application areas that are often common only within a specific field. The set of concepts was then shared with a number of researchers and educators who are involved in research related to network science or who teach related courses. Based on the input provided by these individuals, the committee created the survey of core material in this appendix.

The committee faced a significant challenge as it sought to synthesize the diverse body of material taught in a wide range of disciplines. Indeed, as discussed in Chapter 2, network science is called on to address problems that not only cut across disciplines but also represent a vast body of knowledge, from infrastructure networks, such as the power grid and the Internet, to pervasive applications running over these infrastructure networks, such as the World Wide Web (www); from networks of gene and protein interactions to social and economic networks. The board range of these networks is illustrated in Table C-2.

In what sense can the study of such diverse subjects examples fall under a single unifying domain? The realization that despite their diversity, most real networks are characterized by a small set of organizing principles helped answer this question. For example, the widespread emergence of scale-free networks is rooted in the role of growth and preferential attachment, mechanisms present in many real systems, from cell biology to computer science. This implies that a unified set of tools can be applied to characterize the properties and behavior of a wide range of real networks. For example, tools developed by mathematicians to understand random networks or measures introduced by sociologists to explore social networks can be applied by biologists to design new drugs for disrupting the metabolic network or by computer scientists to explore the properties of the World Wide Web or the Internet. These generic or universal features of real networks and tools are reflected in the courses that are currently taught. Despite their diversity, most courses cover basic concepts that appear to be common across disciplines. The role of this appendix is to survey these common concepts, tools, and methods, identifying the material at the core of network science.

OVERALL ORGANIZATION

Even in established fields of science there is significant room for diversity of focus and interpretation. For example, a survey of basic courses on, say, quantum mechanics or economics would show that while there is broad agreement across instructors, textbooks, and universities on a small set of topics that must be covered, there are significant variations in the examples used, the application areas covered, and special topics. The committee shows that this is eminently the case for network-science-related courses as well. Broadly speaking, one can identify a set of core concepts that emerges in a wide range of courses in network science, largely independent of their discipline. These core concepts are typically embedded into applications with different focus. Applications can be grouped into three main areas. The first and the most dynamically evolving area, biological networks, applies network theory to subcellular (metabolic, regulatory, genetic) networks, neural networks, and ecological networks like food webs and species interactions. The second area, social and economic networks, encompasses a wide range of topics such as social interactions, collabora-

TABLE C-1 Representative Courses on Computer Science

Courses	Institution	Name of Course	Web Address
Core courses	Pennsylvania State University	Graphs and Networks in Systems Biology	http://www.phys.psu.edu/~ralbert/phys597/
	University of Michigan, Ann Arbor	Network Theory	http://www-personal.umich.edu/~mejn/courses/2004/cscs535/index.html
	Columbia University	Networks and Complexity in Social Systems	http://www.columbia.edu/itc/sociology/watts/w3233/
	Columbia University	Scaling in Networks	http://comet.columbia.edu/courses/elen_e9701/2001/outline.html
	University of California at Irvine	Networks and Complexity	http://eclectic.ss.uci.edu/~drwhite/Anthro179a/SocialDynamics02.html
	University of Pennsylvania	Networked Life	http://www.cis.upenn.edu/~mkearns/teaching/NetworkedLife/
	University of Indiana at Bloomington	Structural Data Mining	http://ella.slis.indiana.edu/~katy/L597/
	University of Patras (Greece)	Networks	http://nicomedia.math.upatras.gr/courses/mnets/index_en.html
Applications and related courses	University of Michigan, Ann Arbor	Information Retrieval	http://tangra.si.umich.edu/~radev/650/
	Cornell University	Structure of Information Networks	http://www.cs.cornell.edu/Courses/cs685/2002fa/
	Massachusetts Institute of Technology	Complex Human Networks Reading Group	http://web.media.mit.edu/~tanzeem/cohn/CoHN.htm
	Virginia Tech	Recommender Systems	http://people.cs.vt.edu/~ramakris/Courses/CS6604-RS/outline.html
	Boston College	Social Network Analysis	http://www.analytictech.com/essex/schedule.htm
	University of Toronto	Social Network Analysis	http://www.chass.utoronto.ca/~wellman/courses/gradnet01.htm

tions, social filtering, and economic alliances. The third application area, infrastructure and communication networks, ranges from the Internet and the World Wide Web to power grids, phone networks, and sensor networks.

In the following sections the committee describes in some detail the core concepts, followed by a short discussion of the three application areas. Table C-3 lists some core material that could be expected in network science courses.

NETWORK STRUCTURE

The elementary attributes of all networks are the nodes, which are the basic units of a system, and the links, which are the connections between the nodes. Both the nodes and the links can widely differ in different fields. For example, the nodes might be humans or scientists in social networks; molecules, genes, or neurons in biology; routers or transformers in infrastructural networks; and Web pages or research publications in information networks. Similarly, the links might be friendships, alliances, reactions, synapses, optical and copper cables, URLs, or citations. The totality of the nodes and the links defines a network, often represented in graphic form as a connectivity matrix, telling us which nodes are directly connected to each other. Given that the study of networks was traditionally part of graph theory, a branch of mathematics with long and distinguished history, most of the language that network theory uses today has its roots in graph theory. A network map (or connectivity matrix) is typically the starting point for characterizing the structure or topology of any network.

Once a network has been mapped, the first priority is to characterize its topological and structural features. Degree of connectivity represents the most elementary measure of a node, specifying the number of links a node has to other nodes. Much can be learned about a network by inspecting the degree distribution, which in its simplest manifestation is a histogram of the number of nodes with a given degree. Other important measures include the shortest path between two nodes, which plays a key role in identifying small-world effects; the diameter, which is the distance between the two most distant nodes; the subgraphs and communities that characterize the relationship between small subsets of nodes within a network; the spectral properties, which help us capture a series of local and global network characteristics; and,

TABLE C-2 Real-World Networks Appearing in Courses

Discipline/Course	Type of Network
Infrastructure and communications networks	Power grid Internet Public switched telephone network
Information and content distribution networks	World Wide Web Broadcast Sensors Search
Social networks	Collaboration Communities Social filtering and recommendations Economic Linguistic
Computing networks	Neural nets Petri nets Cellular automata Interacting intelligent agents
Engineering systems	Control networks Integrated circuits Queuing networks Process networks Transportation networks Supply chains and manufacturing
Research networks	Scientific grid Collaborations Blogs and online journals
Military networks	Terrorist networks Intelligence networks Logistics networks
Biological networks	Metabolism Gene and protein interactions Biomanufacturing Regulatory and control networks Ecological networks and food webs Viruses and epidemics

finally, the link strength or weight, which characterizes the nature of the interactions between different nodes.

Based on these measures, real networks may be classified in perhaps two or three major classes. First, there is a class known as regular networks, or graphs, in which the degree of all the nodes assumes the same value or only a few discrete values and the underlying network has a regular, repetitive structure. Such regular graphs approximate the structure of most crystals, as well as a number of other objects, engineered and natural, from the retina of the eye to the roads of some large cities (like New York). Much attention, however, has focused on random networks, systems in which the nodes are randomly connected to each other. In such networks the degrees follows a Poisson distribution. Despite their important role in network theory, we do not know of major real networks that would be fully random. Finally, the availability of large-scale network maps has led to the discovery that many real networks are neither regular nor fully random but, rather, scale-free. They have a heavily tailed degree distribution—that is, there are significant (order of magnitude) differences in the degree of different nodes. Scale-free networks describe the cell, the Web, the Internet, and many collaboration and social and economic networks. While many real networks are intermediate between these three classes, this classification captures some of the basic primitives used in many courses on networks and most networks are characterized in terms of the three classes.

An important question surfacing in many network-science-related courses is the following: What processes and mechanisms give rise to the network characteristics discussed in the preceding section? A closely related question is this: How do we generate networks with structural characteristics that mimic the properties of selected real networks? Network models, introduced to answer these two questions, are an important part of most network science courses (see Table C-4). These models have two main functions. First, some models aim to mimic, in a simplified form, the emergence and evolution of real networks, helping us to understand the mechanism responsible for the formation of real networks. Second, to test the impact of selected network characteristics on the network's behavior, we need to gener-

TABLE C-3 Content of a Typical Network Science Course

Subject	Content
Core concepts	Real-world networks Characterization and classifying networks and their components Network modeling
Network interpretation and processes	Flow and routing Aggregation and growth Communication and coordination
Behavior: networks as dynamic entities	Performance and scaling Robustness Routing and congestion Disruptability
Engineering methods in network science	Network design Network analysis
Applications of network science	Information and communication network Biological networks Social networks Control and mechanical systems Industrial applications Military applications

TABLE C-4 Network Models Commonly Used to Generate Network Topologies and Analytical Tools Used to Characterize and Study the Properties of Models

Network Models	Analytical Tools
Random networks	Exact methods
Erdos-Renyi model	Discrete math
Percolation based	Combinatorics
Scale-free models	Graph theory
Growth and preferential attachment	Dynamical systems
Static models	Master and rate equations
Optimization	Mean field theory
Static models	Generating functions
Small-world model	Stochastic networks
Optimized topologies	Statistical mechanics
	Agent-based models
	Clustering tools

ate synthetic networks with preset properties. Given these needs, network modeling is one of the most highly studied components of network science.

Historically the most studied model is the random network model explored by Erdos and Renyi, which generates networks by placing the links randomly between the nodes. Random networks represent an important reference frame in network modeling. Despite the fact that we do not know of real networks that are captured by this model, given that most of its characteristics can be calculated exactly, it represents an important theoretical tool against which various real networks and hypotheses can be tested. While scale-free network models represent a recent addition to the network literature, given that many real networks of interest, from biology to computer science, have scale-free characteristics, in the past few years they have become the most investigated class of model. Several models are available to generate scale-free networks. The most studied one was introduced by Barabási and Albert (1999) to capture the formation of a real network. It involves two main ingredients, growth and preferential attachment. Given the diversity of networks, a whole class of models has been introduced that incorporate other mechanisms affecting a network's evolution, from fitness to node and link removal and aging. In addition, a number of models do not capture the network formation process but result in scale-free topologies through either a fitness-driven or optimization procedure. Despite the important theoretical role these models play, there is no clear evidence that real networks would be shaped by these processes. The small world model, introduced by Watts and Strogatz (1998), interpolates between regular and random networks and has also generated significant theoretical interest.

Understanding the properties of these networks requires a number of analytical tools that have been developed by a number of fields, from discrete mathematics to statistical mechanics. The study of random networks has a long history, using exact methods developed in graph theory, combinatorics, probability theory, and stochastic processes. Scale-free network models, which represent graphs that change in time, are typically studied using methods based on rate and master equations capable of precisely predicting the degree distribution and other characteristics of scale-free networks. In some cases mathematicians have employed exact methods and the tools of dynamical systems and percolation theory to obtain exact results for these networks. Optimization and genetic algorithms have been used to generate optimized networks. Finally, in order to identify communities and groups in networks there has been a cross-disciplinary interest in developing network clustering methods.

NETWORK DYNAMICS

The specification of the dynamics that characterize a network is less straightforward because these dynamics tend to be rather different in the various applications areas. As mentioned in Chapter 3, the examples range from phase transitions in physical systems, to chemical reactions governed by rate equations, to sociological interactions between people. Here we develop just one example from the computer science discipline.

Computers in a network must often be able to broadcast messages to all other computers in the network. The simplistic protocol would require that each computer maintain a view of the addresses of all computers in the entire network. If computers are constantly leaving or entering the network, the significant problem arises of updating and maintaining a consistent view of the network at each computer. This updating of the views prevented early systems from scaling in size. The solution was for each computer to maintain only a small partial view of the network. The broadcast protocol is for a computer to broadcast to each computer in its view and have computers receiving a broadcast retransmit the message to computers in their views. These more sophisticated protocols that are necessary to overcome scaling problems have lead to sophisticated mathematical techniques and algorithms (Demers et al., 1987). This is just one example of many directions in which basic theory in network science is emerging in the computer science area and appearing in advanced courses.

NETWORK FUNCTION

The purpose of most networks is to transport information, people, or material. These functions often determine both the structure and the dynamics of real networks. Therefore, the understanding of network function has been a very active area in a number of research fields, and it is reflected in most network courses. The high degree of universality and commonality that is present in network structure largely disappears when it comes to network function, thanks to the diversity of the functions that networks assume in different domains. For example, the purpose of the Internet is to transfer information in an efficient manner, in contrast to the purpose of the metabolic network, which is to process the chemicals consumed by the cell and to turn them into the cell's building blocks. Therefore, on the top of similar topologies one can define a wide range of dynamical rules, from flow to diffusion and contagion, that lead to different functions and network behavior. Despite this diversity, a number of common and highly studied themes have emerged in the past few years.

One of the most studied themes focuses on diffusion on networks. Indeed, networks support a wide range of diffusive processes that can cause serious problems in a number of application areas. Social and informational networks are responsible for the diffusion of ideas. Sexual and contact networks are responsible for the spread of infectious diseases and viruses, ranging from acquired immunodeficiency syndrome (AIDS) to influenza to severe acute respiratory syndrome (SARS). Computer and e-mail networks are responsible for the spread of electronic viruses and worms, generating billions of dollars in damages. Finally, social and professional networks are responsible for the spread of innovations, ideas, and rumors, playing a key economic and social role. It turns out that the numerical and analytical tools used by different fields to approach these problems are highly similar. Yet, in the past few years, with the increasing understanding of the topology of sexual contact networks and computer networks, we have witnessed significant paradigm shifts. For example, the discovery that in scale-free networks diseases do not experience an epidemic threshold has had a significant impact on the strategies used for epidemic modeling and on the design of efficient interventions. While these processes are discussed from the perspective of different fields, they are widely covered in a wide range of courses, from biology to business to computer science.

Network flows, describing mostly the situation when something tangible flows from source to destination, are another class of widely studied multidisciplinary problems. They emerge in the Internet as the flow of bits along the physical infrastructure, in the study of metabolic networks as the flux of matter across reactions, or in transportation networks as, for example, the flow of cars on the highway. In general, the network structure canalizes these flow processes and to a high degree determines the flow rates and the necessary capacities on each link and node.

Another much-studied interdisciplinary problem is a network's ability to carry on its functions in the face of errors and failures. Centered on the question of robustness and resilience, many of these studies explore the network's dynamical and topological integrity under node and link loss (Dodds et al., 2003). A number of studies have shown that there is a strong interplay between a network's robustness and its structure. For example, scale-free networks are very robust under random node removal but highly vulnerable to the systematic removal of their hubs. The situation becomes even more complex if network flows are considered, which could lead to cascading failures, such as the 2003 Northeast electricity blackout, affecting millions of consumers (Watts, 2002). But robustness studies play a key role in designing new drugs or in developing systems and network topologies that are highly error resistant.

Another class of much-studied problems focuses on search in networks, leading to algorithms and methods to efficiently locate information in complex networked structures. These studies play a key role in a wide range of problems, from Web search algorithms to the identification of information and expertise in an organization.

REFERENCES

Barabási, A.L., and R. Albert. 1999. Emergence of scaling in random networks. Science 286(5439): 509–512.

Demers, A., D. Greene, C. Hauser, W. Irish, J. Larson, S. Shenker, H. Sturgis, D. Swinehart, and D. Terry. 1987. Epidemic algorithms for replicated database maintenance. In Proceedings of the 6th ACM Symposium on Principles of Distributed Computing. Vancouver, B.C., Canada.

Dodds, P.S., D.J. Watts, and C.F. Sabel. 2003. Information exchange and the robustness of organizational networks. Proceedings of the National Academy of Sciences of the United States of America 100(21): 12516–12521.

Watts, D.J. 2002. A simple model of global cascades on random networks. Proceedings of the National Academy of Sciences of the United States of America 99: 12516–12521.

Watts, D.J., and S. Strogatz. 1998. Collective dynamics of "small-world" networks. Nature 393(6684): 440–442.

D

Questionnaire Data

In this appendix, the committee provides more detail on its analysis of responses to the questionnaire that underlies the results presented in Chapter 6, following the same order of presentation. After describing the questionnaire process and further characterizing the respondents, the committee draws on the responses to more fully characterize the research community's approaches to the possible field of network science: both doubts about the existence of such a field and shared notions of what it would encompass. Although the full context from the corresponding sections of Chapter 6 is not repeated, brief summaries introduce specific details. The appendix also contains the analysis by Katy Börner, of Indiana University, of the social structure of network science revealed by the questionnaire responses.[1]

Network science has reached its present level of visibility from the convergence of two phenomena: the ever-rising importance of networks for the national well-being and security and the achievement of promising formal results (primarily from graph theory) relating aspects of network topology to network properties such as resilience. But long before this convergence, many disciplines studied phenomena arising from recognizable networks in their subject domains. While some of the studies inspired increased multidisciplinary research, their sheer diversity also raised significant doubts about the coherence and usefulness of an underlying "network science" that might investigate further substantive commonalities across the many specific uses of networks as we know them. If there are such commonalities, then there is great potential benefit in pursuing better theories, tools, and insights that address the network science core shared across so many critical domains, and such a pursuit becomes of pressing importance.

The committee adopted an empirical approach to determining the status of this emerging field, if there is one, to assessing a common core for the disparate fields of application, and to identifying the key challenges the science might address. The committee's approach was to solicit the insights of researchers from the communities doing work in the related domains.

THE QUESTIONNAIRE PROCESS

As noted in Chapter 6, the questionnaire text was developed by the committee through an iterative process, starting with the statement of task and continuing through committee discussion and beta-testing. The focus of the committee discussion is clarity and brevity in the questionnaire and consideration of the kinds of information that might reasonably be elicited from the research community. The beta-testing allowed committee members to assess the ease of understanding and responding to the questionnaire text. In its final form, the questionnaire addressed four broad areas: the respondent, the respondent's work, the respondent's view of the potential for network science to exist as a discipline, and an open-ended opportunity to provide further information.

To avoid as far as possible prejudging the question of whether there exists a field of network science—and, if there is such a field, its nature—the questionnaire and the various solicitations of respondents intentionally did not define the term "network science." The committee initially expected that this lack of definition might bring many requests for clarification from the research community; in practice, fewer than 1 percent of those solicited explicitly made such a request. Of those who responded to the questionnaire, 9 percent indicated that the term was unclear or had no well-defined core (see "Dissenting Voices," below). The committee did not determine how many of those solicited decided not to respond because the term was not defined in advance.

Content of the Questionnaire

Box D-1 contains the complete online text of the resulting questionnaire, which was posted as a National Academies of Science (NAS) Web link on December 20, 2004.

---

[1] Katy Börner, associate professor, Indiana University, "Mapping the expertise and social network of network science researchers," briefing to the committee on April 13, 2005.

BOX D-1
NRC Network Science Survey

The National Research Council has commissioned a study of the possibility of identifying "Network Science" as a cross-disciplinary area of research worthy of enhanced attention and funding. The purpose of this communication is to invite you to contribute to this study. You have been contacted because you have been identified as a leader in an area of research that is pertinent to Network Science, and hence as an individual who can contribute significantly to the study as well as benefit from its results and consequences.

The resulting NRC report will be published by the National Academies Press and is expected to influence governmental programs and funding to accelerate and to enlarge the benefits that a focus on Network Science can provide.

We ask you to devote 15 minutes to complete the survey's first two sections. If you have time and interest, please also contribute your insights to the more open-ended questions in sections 3 and 4.

The NRC Network Science Study Committee is grateful for your contribution, and encourages you to invite other researchers to contribute at this study site: http://www8.nationalacademies.org/survey/deps/networksci2.htm.

For further information, please click here, or contact
Dr. Charles Duke, Committee Chair, cduke@crt.xerox.com
Dr. John Hopcroft, Committee Vice-Chair, jeh17@cornell.edu
Bob Love, NRC Study Director, rlove@nas.edu

(1) Your Characteristics
1a. **Name:**
First:
Middle:
Last:
Suffix:

1b. **Contact information:**
E-Mail:
Phone:
Website:

1c. **Principal Fields of Interest** (please mark all that apply)

- ☐ Biochemistry
- ☐ Biology
- ☐ Chemistry (other
- ☐ Organizations theory
- ☐ Physics
- ☐ Political science

than biochemistry)

☐ Computer sciences
☐ Ecology
☐ Economics
☐ Information technology
☐ Internet
☐ Management
☐ Mathematics
☐ Medicine
☐ Operations research
☐ Public health, epidemiology
☐ Public policy
☐ Psychology
☐ Sociology
☐ Telecommunications (other than Internet)
☐ Transportation
☐ Utilities
☐ Others (please specify): [   ]

1d. **Your position or role:** (mark most appropriate one)
[(Click here to choose) ▼]

If other, please describe:
[   ]

1e. **Your organization:**

Name of organization: [   ]

Department: [   ]

Type of organization: (mark one)
[(Click here to choose) ▼]

If other, please describe:
[   ]

(2) Your Work

2a. **Do you consider some or all of your work potentially part of an emerging field of network science?

☐ Yes
☐ No

If yes,

2b. Please briefly describe your particular interests in this area.

[text field]

2c. Please indicate your principal collaborators, using the structure entries below and/or the unstructured text field labeled "Other collaborators".

	Name:	Contact information (website or email):
Collaborator 1		
Collaborator 2		
Collaborator 3		

Other collaborators:

[text field]

2d. Please briefly describe specific projects under which you are pursuing these interests. Use either the structured entries below and/or the unstructured text field labeled "Other projects".

	Project Name:	Project Website:	Principal Investigator(s):	Funding Organization:
Project 1				
Project 2				
Project 3				

Other projects:

2e. **If you have other relevant websites, please list:**

(3) The Field of Network Science

We would appreciate any insight you might give us on the broad questions of defining the proposed field:

3a. **Is there an identifiable field of network science?**
- ☐ Yes
- ☐ No

If yes,

3b. **How would you define it? For example, what are the core topics?**

3c. **What are the driving applications?**

3d. **What are the key research challenges?**

If no,

3e. **What are the principal reasons for your answer?**

3f. Should there be such a field of study?

[text area]

(4) Additional information

We welcome any further information sources you wish to bring to our attention and any material you wish to provide for the Committee's understanding of the field of study. No material will be used in the Report without written permission of the copyright holders.

4a. Other people to whom you suggest we send the survey invitation. Please indicate using the structured entries below and/or the unstructured text field labeled "Other people to invite".

	Name:	Contact information (website or email):
Invitee 1		
Invitee 2		
Invitee 3		

Other people to invite:

[text area]

4b. Pointers to information on network science: books, key papers, websites, conferences, mailing lists, etc.

[text area]

4c. Pointers to other programs funding network science, both US and international:

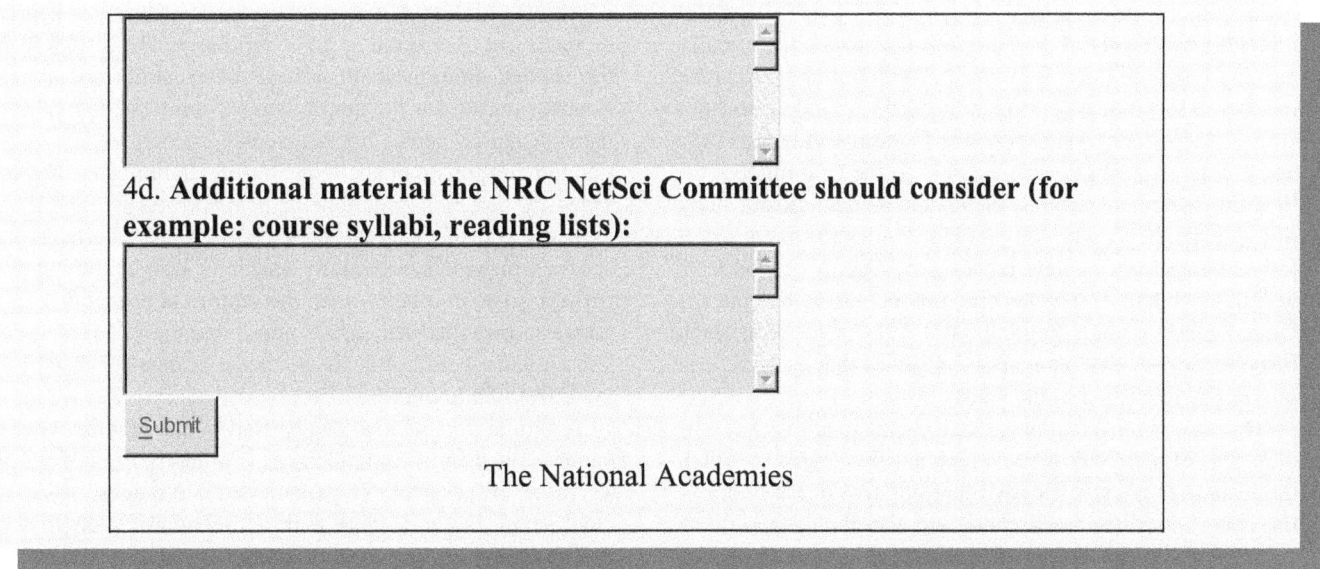

Soliciting Responses to the Questionnaire

The goal of the ensuing solicitation was to reach as large, diverse, and representative a sample of the many relevant research communities as feasible within the study's resources. Because both the full range of relevant communities and the populations of researchers within those communities could not be readily defined in advance, the primary process of choice was a snowballing outreach. To begin the process, 113 recognized researchers working on topics in candidate areas clearly relevant to the possible science of networks were sent e-mails asking them to complete the questionnaire. These solicitations briefly described the nature and purpose of the study, provided the URL (Uniform Resource Locator) of the on-line questionnaire, and invited the recipient to forward the announcement to other researchers who might be interested. The solicitation process was then continuously iterated, drawing on the responses to identify more people—collaborators, project principal investigators, and so on—to receive a solicitation. The snowballing process was stopped whenever it stepped outside the field of network science as indicated by the respondents—that is, if a respondent explicitly indicated that he or she was not working in "network science," the respondent's list of collaborators and principal investigators was not added to the pool of names. As of April 29, 2005, a total of 2,040 people had been directly contacted through the solicitation process.

Some heuristic sanity checks were performed to catch hoax entries; only three of the responses appeared suspect, and their content did not affect the conclusions of this study. A variety of spot checks were performed. For example, Is the response internally consistent? Does it appear to name nonexistent or wildly irrelevant people, programs, or organizations? Does a respondent appear as an author in the technical literature (as indexed by Google Scholar Beta). However, such checks were limited and are indicative at best; a thorough screening analysis was not attempted.

There are many inherent limitations to a snowballing process. For instance, poorly connected members of the underlying communities might be left out, or there could be under- or overrepresentation of communities or specific programs owing to differences in willingness to respond to such a questionnaire or to provide information that would allow further snowballing. In addition, the committee observed that a few highly connected people provided no information on collaborators or projects: Instead of listing names and projects, they sent replies such as "too many to list." The committee was especially concerned about these outreach limitations, as the on-going questionnaire analysis quickly demonstrated that many target communities were weakly interacting.

The committee followed several ancillary processes to offset the limitations of snowball instituted coverage by bringing in additional sources of names throughout the solicitation process: literature citation studies, sequential tracing of collaborative ventures, conference attendance, mailing lists, and personal interviews with the authors of recent books and reviews.

The citation study and analysis looked at some selected researchers' work and collected the names of coauthors, cited authors, and authors who cited those researchers' work. A key goal of this analysis was to improve the coverage of subject fields that had not yet seen many questionnaire responses. Although resource limitations constrained the amount of citation analysis that could be performed, the amount that was done succeeded in introducing several hundred names that had not been uncovered by the snowballing process to that point. Spot checks suggest that at least several thousand additional names might have been produced by more intensive citation tracking.

In view of the methodological concern over potentially uneven community representation due to systematically varying response rates, it might be worth noting that (1) the number of new names provided by each respondent was not strongly dependent on his or her self-identified field of study and (2) the overall rates of response to the committee's solicitations were not strongly dependent on the fields of study of the respondents who provided the names to solicit. Similarly, the number of new names provided by respondents was independent of whether the respondents were from the United States or not. Regardless of field or location, each respondent provided a mean of 2.8 names that had not been previously identified in the study.

The committee recognizes that it cannot quantify the completeness of the resulting coverage nor the degree to which the responses are statistically representative of the underlying communities of researchers (see also the discussion of coverage saturation below). For this reason, some classes of analysis could not be reliably performed and are not addressed in this report: For example, the committee explicitly chose not to attempt to identify a top-100 list of researchers, programs, or institutions. Nonetheless, its analysis of the key responses relating to the existence and nature of a possible field of network science appears solid: The responses are stable across all responses obtained when they are partitioned by such factors as when in the solicitation process the response was received, whether the response was directly solicited or not, whether a solicitation was generated by snowballing or from the ancillary sources, which research communities the respondents self-identified as their own, and what country the respondents worked in. Some distinct differences appear between respondents who believe there is a field of network science and those who do not; these differences are described below.

The committee is confident that the solicitation process, despite the multiple approaches and continued effort, did not saturate the population of researchers whose work touches on the potential field of network science. A variety of heuristic measures contribute to this confidence. The results from the limited citation analysis have been mentioned above. Another reason for the committee's confidence is that new names (that is, names not previously encountered in questionnaire responses or ancillary sources) continued to be provided by successive increments of responses without letup until the end of the study (see Figure D-1). The latest responses provided essentially as many new names as the earliest ones; in other words, the empirical probability of a response-provided name falling outside the set of already-known names did not decrease as the number of responses grew from 50 to over 600.

A similar conclusion is suggested by the fact that once a name is cited by a respondent, it is unlikely to be cited by any other respondent: 71 percent of cited names are never cited again. In short, there is no indication of saturation in the coverage, so one may conclude that the questionnaire solicitation process does not approximate complete coverage of those who would be interested.

THE RESPONDENTS

Over half of the responses (57 percent) came from people who had been directly invited during the snowballing process; the remaining spontaneous responses are believed to have largely been induced by individuals forwarding the solicitation note and by its dissemination in online mailing lists (see Figure D-1). The questionnaire did not ask how the respondent learned of the study, but this substantial proportion of spontaneous responses sheds light on both the limitations of the committee's explicit snowballing and the effectiveness of using additional solicitation mechanisms. In aggregate, names of 2,374 distinct people were provided by these responses and the ancillary sources, although valid e-mail addresses were identified for only 2,123 of them.

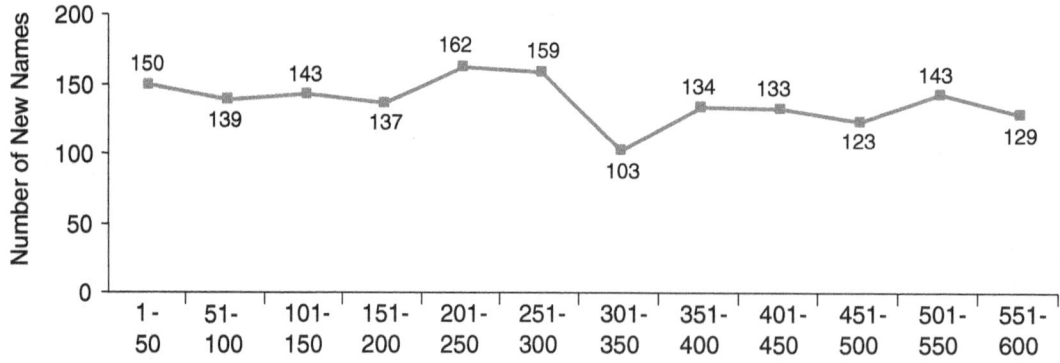

FIGURE D-1 New names by response ID.

Geographic Locales

Questionnaire responses were received from 29 countries; the two most recurrent were the United States (497 responses representing 39 states) and Canada (23 responses representing 6 provinces) (see Figure D-2).

In analyzing each question, the results for individual countries were compared against the aggregate results (see Tables D-1 and D-2). Because most countries had few entries, U.S. responses were compared with the aggregate figures for all non-U.S. responses. No significant differences appeared. For example, the percentages of those self-identifying their work as being in network science, of those stating there is an identifiable field of network science, and of those providing definitions, interests, application, and challenges were closely comparable. Similarly, there were no significant differences in the mean number per response of fields selected; collaborators; projects; or new names (or of new names that later responded). This was also true when U.S. responses were analyzed by state (see Figure D-3 and Table D-3).

Fields of Study

The best-represented fields, as identified by the respondents, are computer science (and its closely related areas), other (described in more detail below), math, biology, and physics (see Figure D-4).

The questionnaire was structured to allow each respondent to indicate more than one field of interest, and this opportunity was heavily used: the mean number of fields selected by a respondent was 3.6, and 80 percent of the respondents selected more than one field. For this reason, the response-per-field figures shown in Table D-4 are 3.6 times the number of responses.

These data also demonstrate the success of the solicitation process in bringing in research communities that had not been identified in advance as involved in network science: Some 159 (28 percent) of the responses indicated a field other than the fields initially provided by the online questionnaire (see Figure D-5). Analysis of the free-form text entries describing these other fields shows great diversity, with the most numerous being engineering, geosciences, and human communication. The category labeled "Unclassified other" represents fields with single entries; examples include botany and economic history.

The respondents overwhelmingly came from academia; of the 619 respondents (98 percent of all responses) who indicated the type of organization they worked in, the questionnaire received only 46 (7 percent) from industry and only 12 (2 percent) from the military (see Table D-5). One significant contributor to the low representation from industry and the military was the comparative difficulty of finding e-mail addresses or other contact information for people outside academia. Compounding this problem, the effectiveness

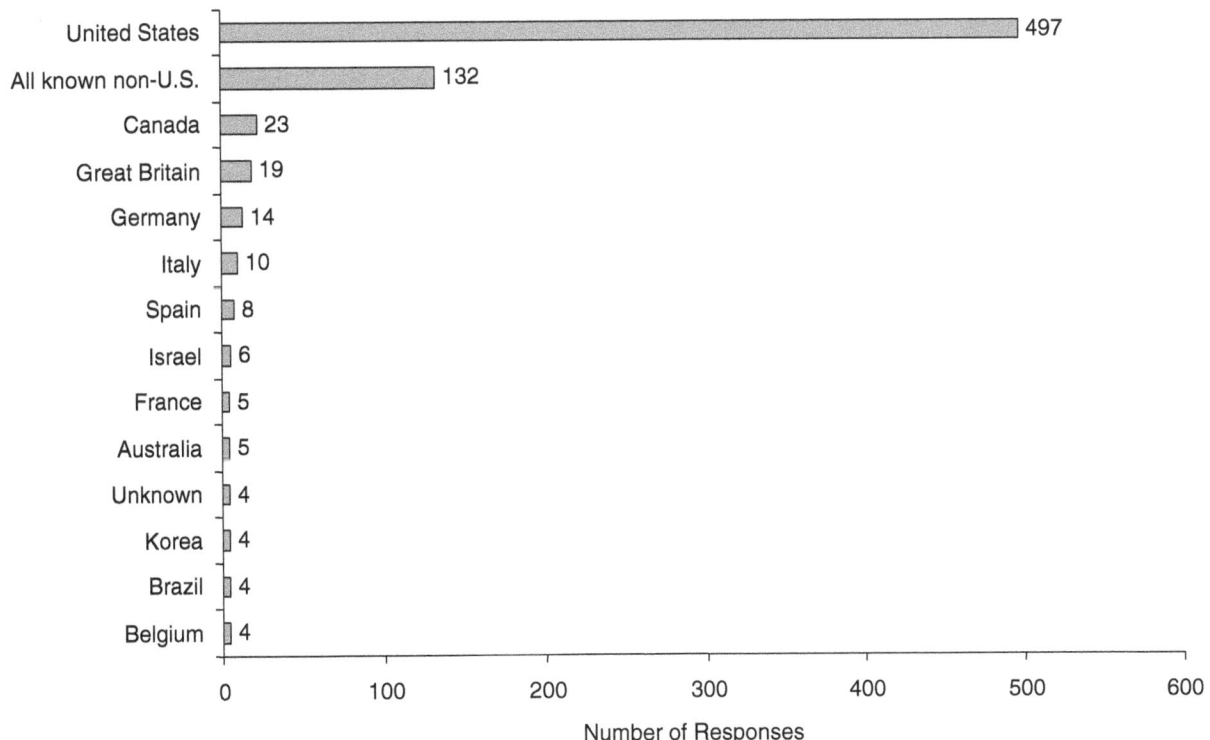

FIGURE D-2 Countries where respondents were located.

TABLE D-1 Respondent's Country

Country	Number of Respondents Who Selected the Country	Percent of Respondents Who Selected the Country
Known non-U.S.	132	20.9
Argentina	1	0.2
Australia	5	0.8
Belgium	4	0.6
Brazil	4	0.6
Bulgaria	1	0.2
Canada	23	3.6
China	2	0.3
Denmark	3	0.5
France	5	0.8
Germany	14	2.2
Great Britain	19	3.0
Greece	1	0.2
Hungary	1	0.2
India	2	0.3
Israel	6	0.9
Italy	10	1.6
Japan	3	0.5
Korea	4	0.6
Mexico	1	0.2
Netherlands	2	0.3
New Zealand	1	0.2
Poland	1	0.2
Portugal	2	0.3
Russia	2	0.3
South Africa	1	0.2
Spain	8	1.3
Sweden	3	0.5
Switzerland	3	0.5
United States	497	78.5
Unknown	4	0.6

TABLE D-2 Canadian Respondent Provinces

Province	Number of Respondents Who Selected the Province	Percent of All Respondents Who Selected the Province	Percent of Respondents in Canada Who Selected the Province
Alberta	1	0.2	4.3
British Columbia	7	1.1	30.4
Newfoundland	3	0.5	13.0
Nova Scotia	3	0.5	13.0
Ontario	8	1.3	34.8
Quebec	1	0.2	4.3

of snowballing was significantly greater for academic respondents: On average, each response from academia provided 3.1 new names, while responses from outside academia provided only 2.1 new names. In turn, the people identified by academic respondents were also 50 percent more likely to respond. The committee speculates that researchers in academia may perceive more incentive to respond and may attach more importance to influencing the study. Judging from personal experience and anecdotal evidence, industrial researchers today are under intense pressure to focus on near-term financial return.

The respondents were nearly unanimous in describing their own work as related to network science: Only 24 respondents (4 percent) did not so describe it. An additional 1 percent of the solicitations elicited personal e-mails to the committee indicating that the recipient declined to submit the questionnaire, usually because he or she did not work in the area (see Table D-6). The near unanimity on working in network science was independent of the specific fields that researchers worked in and of the country where they worked. Ninety-seven percent of respondents from academia said they worked in network science, as did 93 percent of the other respondents. Overall, this result is a reminder that the results of the questionnaire reflect self-selection on the part of those who responded; it must also be considered in light of the fact that only 70 percent of these respondents indicated that there was an identifiable field of network science (see "Dissenting Voices," below).

DISSENTING VOICES

The questionnaire analysis demonstrates that there is a widespread but not universal belief among the respondents that there is an identifiable field of network science. Although 95 percent classify their own work as potentially belonging to an emerging field of network science, only 70 percent state that such a field is currently identifiable. The main reasons for saying there is no such field are that the term has no coherent definition, that it is broad to the point of vacuity, that it is too soon to define the field, that the field is merely a new name for an already existing field, or that it represents the wrong approach.

The lack of consensus is shown clearly in the questionnaire responses: Of the responses that had been received as of April 29, 2005, only 442 (70 percent) answered yes to Q3a: Is there an identifiable field of network science? Of the remaining responses, 146 (23 percent) answered no and 45 (7 percent) did not answer. These percentages proved stable as the number of responses grew and were only mildly dependent on the field of study of a respondent. Table D-7 lists the fields in decreasing order of the proportion of positive responses to the question. Only political science and public policy are notably more skeptical. Many more fields have yes percentages above the mean for the entire sample than below it: This "Lake Wobegon" effect primarily arises from a positive correlation between answering yes to this question and marking oneself in more fields of study.

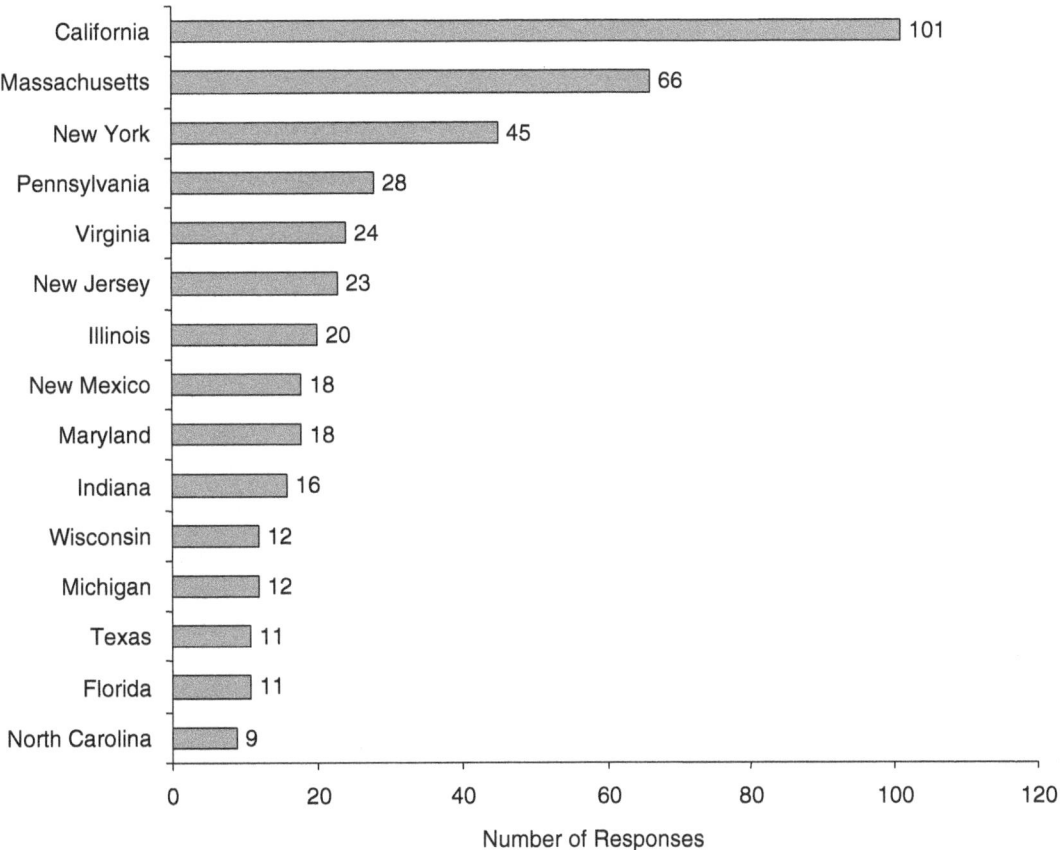

FIGURE D-3 States where respondents were located.

Although 23 percent of the respondents explicitly stated that there was no such identifiable field, there is some sign in the results that more than 23 percent felt that the question was debatable: 26 percent answered Q3e, "If no, what are the principal reasons for your answer?" and 33 percent answered Q3f, "Should there be such a field of study?"

Q3e allowed free-form descriptions of the principal reason for saying there was no identifiable field of network science. The committee analysis of the 163 responses to this question indicates five broad reasons; a given respondent often would offer more than one reason (see Figure D-6). In addition, respondents also said that the field suffered from excessive hype.

The five broad reasons there is no field of network science are these:

- The term is unclear or has no coherent core. For example, "Network science combines two words such that the resulting pair specifies less information than either individual word alone."
- The term reflects a field that is still emerging, so it is too early to tell if it will bear substantive results. This phrasing may occasionally be a more tactful variation on the preceding wording, but often specific emerging application domains are mentioned.
- The work labeled network science is simply part of some specific existing fields under a new name. There is disagreement on just what existing field it is; frequent candidates include graph theory, complexity theory, systems theory, computer science, and control theory.
- The phrase is too broad, to the point of being vacuous: anything can be represented as a network, but doing so does not provide meaningful insight. For example, "Network theories turn out wrong when applied to particular application areas."
- The idea of developing network science as a separate field is a wrong or barren approach. For example, "The interesting questions arise from function, rather than topology." These answers agree that there is work that can be called network science but disagree that developing it as a discipline science will benefit the many other application domains that refer to networks in some form. When these comments indicate what the

TABLE D-3 Respondent States

State	Number of Respondents Who Selected the State	Percent of All Respondents Who Selected the State	Percent of Respondents in the U.S. Who Selected the State
Alabama	2	0.3	0.4
Arizona	7	1.1	1.4
California	101	16.0	20.3
Colorado	8	1.3	1.6
Connecticut	2	0.3	0.4
Delaware	1	0.2	0.2
District of Columbia	3	0.5	0.6
Florida	11	1.7	2.2
Georgia	6	0.9	1.2
Hawaii	6	0.9	1.2
Illinois	20	3.2	4.0
Indiana	16	2.5	3.2
Iowa	2	0.3	0.4
Kentucky	4	0.6	0.8
Louisiana	1	0.2	0.2
Maryland	18	2.8	3.6
Massachusetts	66	10.4	13.3
Michigan	12	1.9	2.4
Minnesota	7	1.1	1.4
Mississippi	1	0.2	0.2
Missouri	5	0.8	1.0
New Hampshire	1	0.2	0.2
New Jersey	23	3.6	4.6
New Mexico	18	2.8	3.6
New York	45	7.1	9.1
North Carolina	9	1.4	1.8
North Dakota	1	0.2	0.2
Ohio	6	0.9	1.2
Oklahoma	1	0.2	0.2
Oregon	3	0.5	0.6
Pennsylvania	28	4.4	5.6
Rhode Island	2	0.3	0.4
South Carolina	2	0.3	0.4
Tennessee	5	0.8	1.0
Texas	11	1.7	2.2
Vermont	2	0.3	0.4
Virginia	24	3.8	4.8
Washington	5	0.8	1.0
Wisconsin	12	1.9	2.4

"wrong approach" is, they often mention the use of connectivity or topology alone as defining the scope of network science.

The responses also indicate that skepticism about the existence of network science is encouraged by a perception that the term "network science" has become trendy, both overstated in its claims and overused as a justification for other work. One explicit aspect in the expressions of perceived over claiming and overuse is that if network science is defined solely in terms of "anything described as connected," there is little that can be excluded from the claimed realm of application.

DEFINING THE FIELD

The first question in proposing a possible discipline of "network science" is this: What are we studying? As has already been seen in this report, many domains use the word "network," but what are the defining characteristics that let us recognize a network as a subject for study in "network science"?

Two questions on the questionnaire are directly designed to address this issue: question 3a, which asks whether there is a defined field of network science, and question 3b, which asks those who responded yes to 3a to define the field as they see it. The 436 responses received through April 29, 2005, to these specific questions (out of 633 total responses) give the committee empirical data on the nature of "network science" as practiced by current researchers in the various domains. In addition, the responses to four other questions proved highly relevant to this subsection and were reviewed for consistency with the conclusions presented here: question 2b, which asks the respondent to describe his or her current research interests (596 responses); question 2d, which addresses specific current research projects (436 responses); question 3c, which asks about the driving applications of network science (427 responses); and question 3d, which asks about research challenges in the area (424 responses). These results must also be viewed taking into account that 30 percent of respondents did not believe such a field is currently identifiable; their concerns are discussed above.

The many definitions and research interests provided by the research community did allow formulating a potential core definition of network science. To organize its description of this core, the committee structured its analysis in terms of two basic components that identify any field of study; this initial organization was chosen to be high level, to focus on the information that was needed without prejudging what might actually be found in the questionnaire data:

- *The subject being studied.* If a well-defined field of "network science" is to exist, the defining attributes of a "network" must be determined. A given network or class of networks can then be characterized by specific input values of these attributes. An attribute may have very complex or dynamic values, defined by state structures, algorithms, models, or empirical data.
- *The derived properties of interest.* When one solves a problem in "network science," what does one want to know? What derived properties or insights arise from the input attributes of the networks being studied? Here again, a proposed network science requires that these kinds of output properties can be meaningfully identified across the range of application domains.

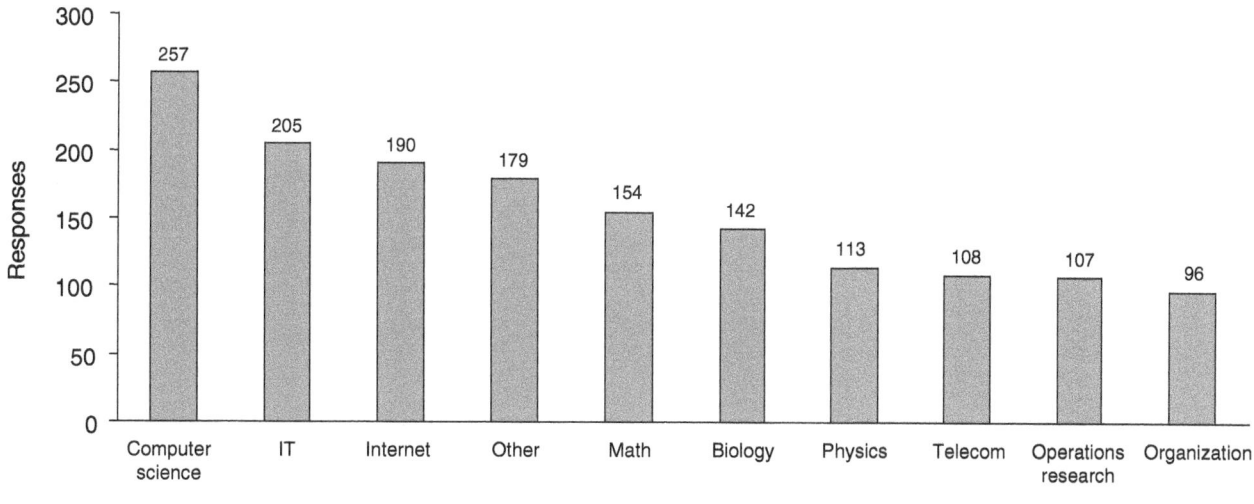

FIGURE D-4 Fields selected by respondents.

TABLE D-4 Responses per Field

Field	Number of Respondents Who Selected	Percent of Respondents Who Selected	Mean Number of Fields Selected
Computer science	232	40.8	4.82
Information technology	186	32.7	5.28
Internet	178	31.3	5.32
Other	159	28.0	3.91
Math	137	24.1	5.51
Biology	136	23.9	4.51
Physics	103	18.1	4.88
Telecom	98	17.3	5.50
Operations research	90	15.8	5.90
Organization theory	86	15.1	6.21
Sociology	80	14.1	5.68
Ecology	75	13.2	5.53
Economics	74	13.0	6.92
Management	69	12.1	6.42
Public policy	61	10.7	6.79
Biochemistry	59	10.4	5.27
Political science	47	8.3	6.15
Medicine	44	7.7	5.75
Public health	42	7.4	6.31
Psychology	36	6.3	6.97
Transportation	34	6.0	6.35
Chemistry	22	3.9	6.73
Utilities	18	3.2	6.44
Overall respondents			
Total	2,066	363.4	
Mean	89.83	15.8	3.64
Median	75	13.2	3.00

If these questions have common answers across many application domains, then network science might then be identified as the insights, lexicon, measurements, theories, tools, techniques, and heuristics that allow one to map between the desired output properties of given networks and their input attributes. Mapping is needed in both directions: determining the output properties that arise from specific input attributes, and determining the input attributes that could be designed into a new network or achieved by intervening in an existing network to realize particular output properties. If network science is to meaningfully be said to exist, these techniques must be effective throughout many application domains, with well-understood means to apply the general methods to specific domains. For a hypothetical example, one might envision a simulation tool that efficiently dealt with network models across a wide range of size and timescales, with a growing suite of model libraries customized to specific application domains: ecological networks, metabolic networks, transportation networks, and so on.

Just as the fundamental concepts of structure and dynamics shape the committee's overall discussion, they also provide the key to decomposing the inputs, problem dimensions, and outputs of network science into specific roles within a given research or engineering study. Structure and dynamics are orthogonal abstractions applicable to each of these factors: Indeed, a given input attribute, problem dimension, or output property may contribute to both the structural and dynamic aspects of a given study, depending on the focus and intent of the study.

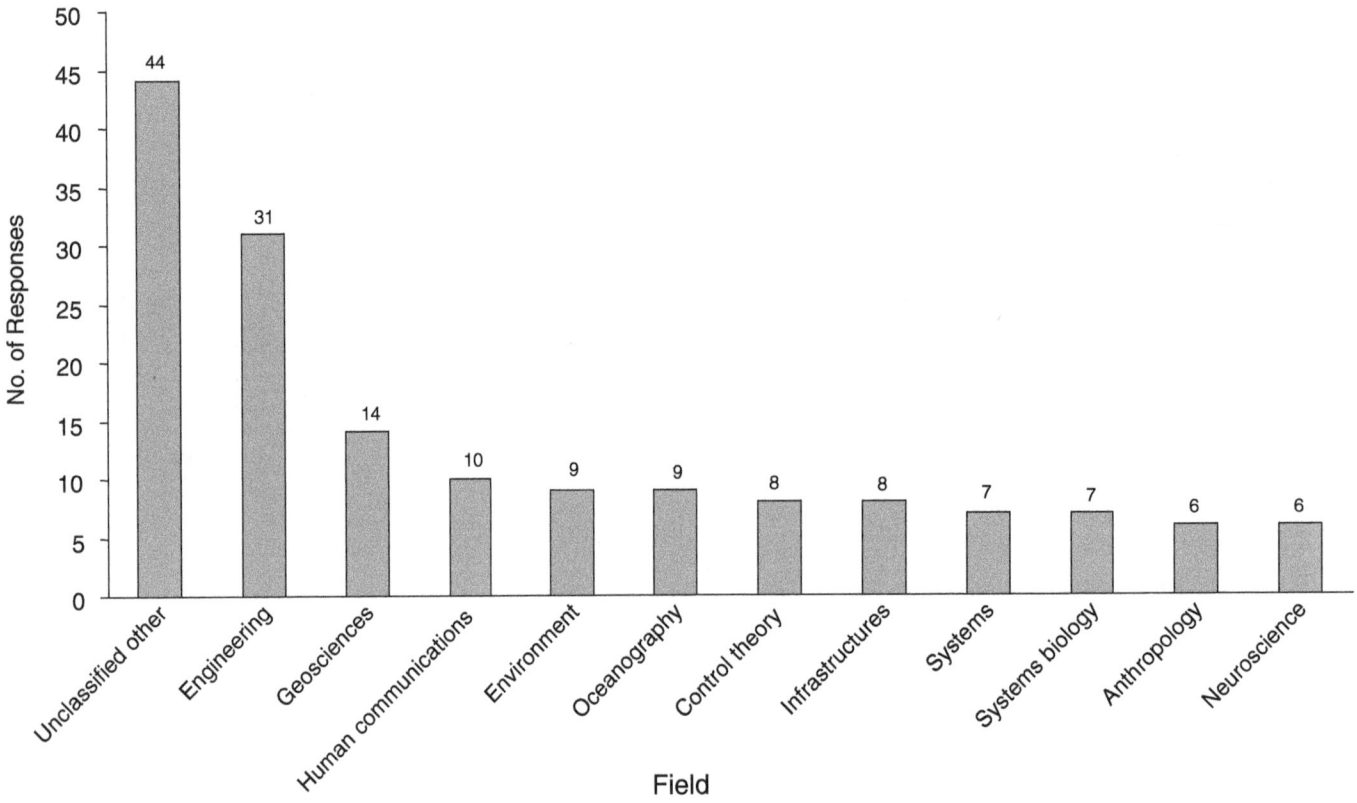

FIGURE D-5 Most frequently mentioned fields.

The reader may find it useful to supplement the somewhat abstract definitions in the following subsection with the hypothetical description of a highly simplified networking problem given further on.

Defining Attributes of a Network

As noted in Chapter 6, analysis of the responses reveals three common attributes of networks: (1) they consist of nodes connected by links; (2) nodes exchange resources across the links; (3) nodes only interact through direct linkage; for brevity, these attributes are designated "connectiv-

TABLE D-5 Respondent Affiliations

Organization Type	Responses
University	455
Other nonprofit	38
Industry	46
Private consultant	18
Military	12
Other governmental	30
Other	20
TOTAL	619

TABLE D-6 Is Your Work Potentially Part of an Emerging Field of Network Science?

	Yes		No		Didn't Say	
Field	No.	%	No.	%	No.	%
Biochemistry	61	100	0	0	0	0
Biology	141	99	1	1	0	0
Chemistry	25	100	0	0	0	0
Computer science	250	97	6	2	1	0
Ecology	75	95	4	5	0	0
Economics	83	99	1	1	0	0
Internet	183	96	7	4	0	0
Information technology	196	96	9	4	0	0
Management	74	94	5	6	0	0
Math	149	97	4	3	1	1
Medicine	44	98	1	2	0	0
Operations research	106	99	1	1	0	0
Organization theory	94	98	1	1	1	1
Other	169	94	9	5	1	1
Physics	110	97	3	3	0	0
Political science	51	98	0	0	1	2
Psychology	38	97	1	3	0	0
Public health	43	100	0	0	0	0
Public policy	69	99	1	1	0	0
Sociology	81	96	2	2	1	1
Telecommunications	107	99	1	1	0	0
Transportation	41	98	1	2	0	0
Utilities	21	91	2	9	0	0

TABLE D-7 Is There an Identifiable Field of Network Science?

Field	Yes No.	Yes %	No No.	No %	Didn't Say No.	Didn't Say %
Utilities	20	87	2	9	1	4
Physics	92	81	18	16	3	3
Public health	34	79	7	16	2	5
Transportation	33	79	8	19	1	2
Medicine	35	78	8	18	2	4
Biology	110	77	24	17	8	6
Ecology	60	76	15	19	4	5
Biochemistry	46	75	13	21	2	3
Telecommunications	81	75	20	19	7	6
Math	115	75	33	21	6	4
Information technology	152	74	41	20	12	6
Economics	62	74	18	21	4	5
Internet	138	73	38	20	14	7
Chemistry	18	72	7	28	0	0
Psychology	28	72	9	23	2	5
Computer science	184	71	61	24	12	5
Sociology	59	70	19	23	6	7
Operations research	75	70	27	25	5	5
Other	124	69	41	23	14	8
Organization theory	66	69	26	27	4	4
Management	53	67	20	25	6	8
Political science	31	60	17	33	4	8
Public policy	40	57	24	34	6	9

ity," "exchange," and "locality." Table D-8 summarizes the structural and dynamic composition of these attributes.

Few responses simultaneously capture all three attributes, but across a wide range of subject domains, these three attributes consistently appear explicitly or by implication in more domain-specific entries. The proportion of responses driving the committee's identification of each input attribute is shown in Figure D-7. Because all three attributes are inherent to defining and understanding a network, the discussion of each attribute has frequent cross-references to the other attributes.

Connectivity

A network has a well-defined connection topology, in which each discrete entity ("node" in graph-theoretic terminology) has a finite number of defined connections ("links") to other nodes. A given link is commonly, but not necessarily, point to point: that is, it connects two nodes. Multipoint links, where they exist, may often be adequately modeled as collections of point-to-point links. A given link may or may not have a defined direction ("from node A to node B"); undirected links may generally be modeled as pairs of directed links. The set of links associated with a given node may change over time, but at any given moment a node has only a finite number of nodes to which it is linked (its "neighbors"). The nodes and links are defined by both structural attributes and dynamic attributes. The structural attributes include the current snapshot of the underlying graph: which nodes are linked to which others. Structural attributes also

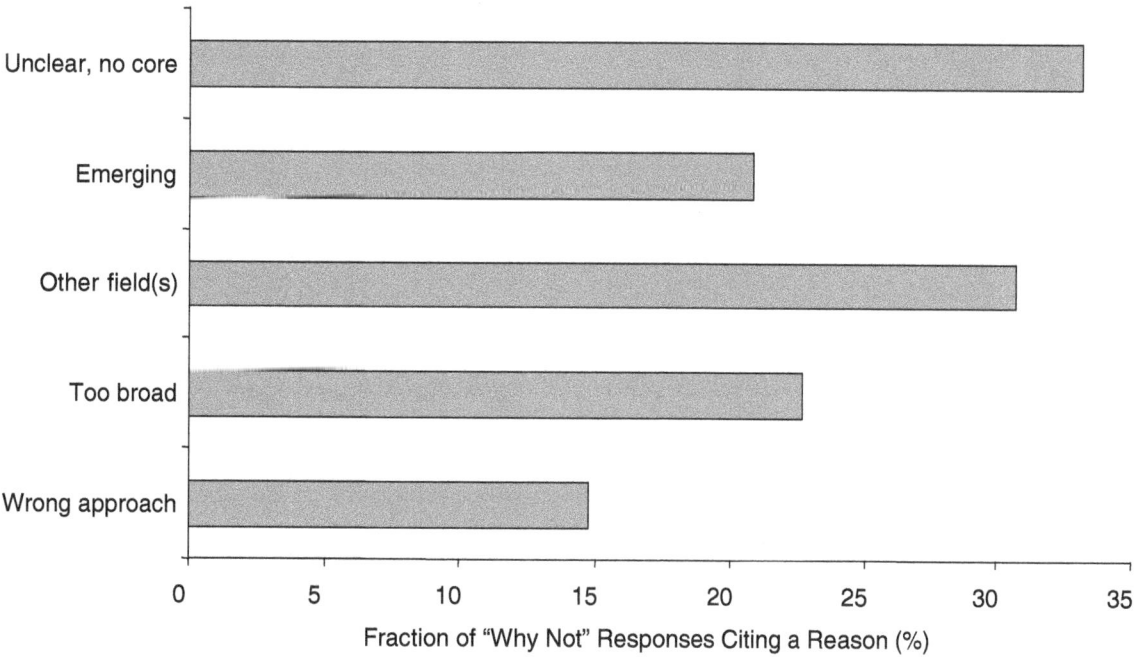

FIGURE D-6 Reasons for saying there is no field of network science.

TABLE D-8 Summary Decomposition of the Input Attributes of Networks

Attribute	Structure	Dynamics
Connectivity—nodes, links, and their specifications	The structural constraints and permissible states of nodes and links. The snapshot at a given instant of the nodes, links, and their attributes (the values of their structural state variables). Note that locality implies that individual nodes cannot have a consistent view of the complete network structure unless the structure changes slowly relative to exchange.	The state transition models of the nodes and links that define their behavior. The mechanisms for the addition, modification, and deletion of nodes, links, and their specifications.
Exchange—resources and their transport, storage, and transformation	Resource classes, specification of the mechanisms and resources that realize resource transformation.	Storage and transport capacities. Interactions between resource exchanges. Latencies in transport and transformation. Degradation of resources during transport or storage.
Locality—node behavior and state; interactions as exchanges over links	The required capabilities that must be installed in nodes and links to support their local behavior.	The definition of the local behavioral rules that govern interactions between agents.

include exchange-related attributes, such as capacity. The dynamic attributes include evolution of the underlying graph (addition and deletion of nodes and links) and the resource exchange attributes of these nodes such as transit time across links and resource transformation mechanisms.

Exchange

The connection topology exists in order to transport one or more classes of resource between nodes; indeed, a link is represented as existing between two nodes if and only if resources of significance to the network domain can be directly transported from one of the nodes to the other without the intervention of other nodes along the transport path. An instance of a network is then characterized by the classes of resource that constitute its exchanged payload of interest. To make this abstract characterization concrete, some examples of resources exchanged in specific networks include bits in computer nets; cargo for bulk transportation; people for passenger flights; energy for chemical reaction networks; and influence for social relations.

The same system can be viewed as a different network depending on which resource exchanges are of interest, so that, for example, the same system might be analyzed as an electrical power distribution network when the resource of interest is electricity and as a telecommunications network when the resource is information encoded as bits distributed over the electric lines. A given network may carry multiple classes of resource, whose differences are reflected in the constraint models that characterize and interrelate the network's attributes (see below). In defining the exchanges of a given network, one must define such characteristics, whether a resource exchange is logically continuous (a stream) or in discrete independent units, how the resources are transported over the links and stored at the nodes, and what resource transformations are performed at nodes or

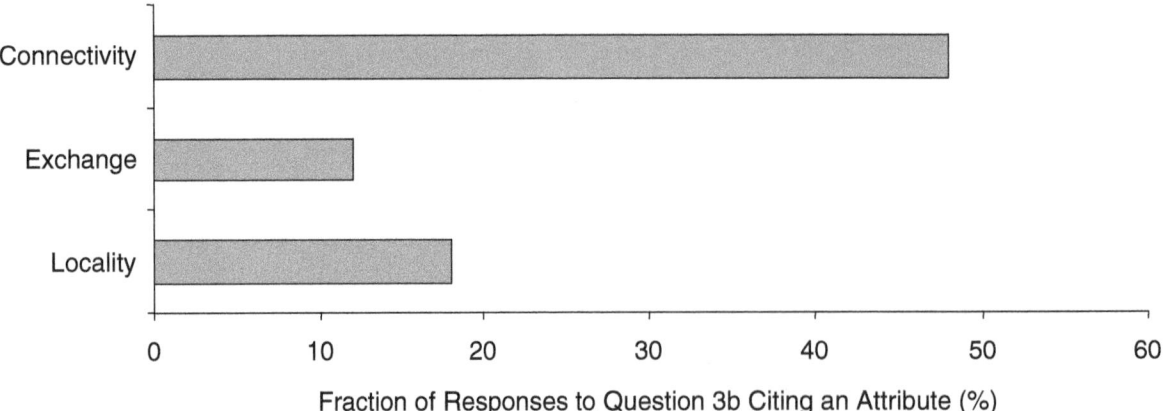

FIGURE D-7 Responses identifying network attributes.

during transit over links. Technically, there is also short-term storage on the links, because transport speeds are finite.

Locality

The exchanged resources are delivered, and their effects take place, only in local interactions (node to link, link to node). There is no God's-eye view of effects and resources, and the effects take time to propagate. This locality of interaction implies autonomous agents acting on locally available state. (When focusing on locality issues, the term "agent" may therefore be used instead of "node.") Like the other key attributes of a network, locality appears in both the structural and behavioral components of a network's definition. Each node and link includes in its structure its available resources, its individual goals (which may be modeled as local values for the cost models and benefit models discussed below), and the mechanisms available to it for achieving them. The dynamic components of a network's locality definition can be expressed as algorithms for maintaining local state and exchanging resources.

In particular, all networks reflect the dynamic consequences of locality, yielding phenomena that appear across the many application domains of network science, such as wavefront effects in the spread of resources and feedback and stability issues due to control delays. A fundamental consequence of locality is that globally optimal structure and behavior are hostage to the independent local optimizations of the individual nodes and links. The aggregate network may become trapped in an equilibrium in which the system is maintained in a nonoptimal state by the independent optimization behaviors of its individual components. Locality often entails an analysis approach that is essentially game-theoretic.

Derived Properties of Networks

Analysis of the proposed definitions in the questionnaire responses also identified six derived properties that spanned a wide range of application domains: characterization, cost, efficiency, evolution, resilience, and scalability (see Figure D-8). The responses driving the committee's identification of the common derived properties of interest in network science are tabulated in Table D-9.

Shared Aspects of the Network Science Problem Space

Beyond these shared input attributes and derived properties of interest to problems in network science, several ancillary concerns were shared across many application domains; these may be broadly classified as constraint models and problem dimensions.

Constraint Models

The network attributes are tied together by cost and benefit models that define the mapping from a network's links, nodes, exchange model, and local agents to derived penalty and merit figures. The network's creation, operation, and behavior, as well as its growth, repair, and evolution, are

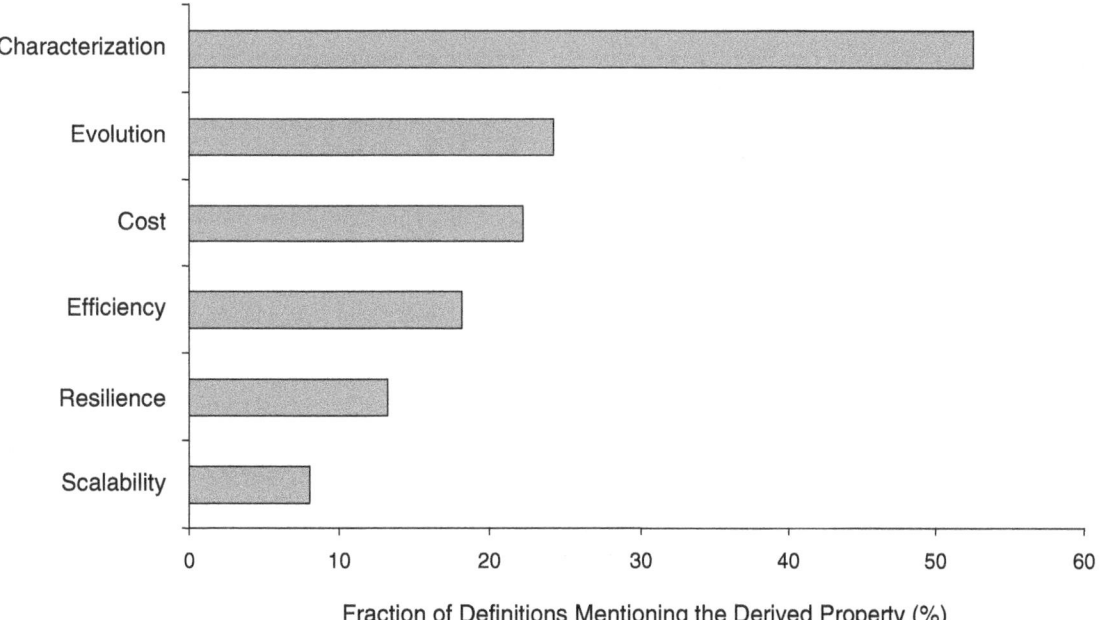

FIGURE D-8 Derived properties of networks mentioned by respondents.

TABLE D-9 Summary Decomposition of the Derived Properties of Networks

Property	Structure	Dynamics
Characterization	Which nodes and links are "important"? What roles do they play in the network? How would one modify a network to change the role of a node in a specified way?	What is the performance of the network, typically in terms of qualities of the resource exchange? What network attributes are required to achieve specified behavior? How do we measure performance and maintain it?
Cost	What is the cost of the aggregate suite of nodes and links of a network, given their defined attributes?	What is the cost of the latency and degradation of the resource exchange realized by this network? What is the least cost network, given input constraints, to achieve a specified performance?
Efficiency	What is the expected utilization of the network nodes, links, and their limiting resources?	What is the trade-off between cost and performance for the available design space? How could the behavioral attributes of the network components be modified to improve the efficiency of the network realization?
Evolution	Which structural attributes are preserved as the network evolves? What structure should be created to assure its stable evolution?	What evolutionary path will emerge under specific rules and constraint models for the addition, modification, and deletion of nodes, links, and their attributes? How can one design or promote local behaviors that will result in a desired evolutionary path? How do failure modes and attacks evolve in response to the network evolution?
Resilience	What are the structural attributes that resist accidental or intelligently planned damage and overload?	How does the behavior of a network change under damage and overload? What input behavioral rules induce better behavior under these scenarios?
Scalability	Which structures scale in terms of the measures of complexity?	How does network behavior change as network scale changes? What rules and constraint models assure desired behavior across changes in scale?

driven by these defining constraint models through feedback or signaling mechanisms.

The cost models and benefit models may have both structural components (determined by static attributes of the network) and dynamic components. "Cost" is an abstract term measuring consumption of resources or decrease in value; while engineered networks may have cost models that output actual dollar costs (among other penalty factors), many networks measure costs in other units. The dynamic components of a cost model reflect aspects of the temporal behavior of the network connectivity and exchange, including the degradation of an exchanged resource's value and the consumption of resources on the nodes and links. As mentioned above, both nodes and links have dynamic components to their definition in that they transform resources passing through them. The transformation is reflected in both cost models and benefit models defining the network. Depending on a given network's definition, a transformation may contribute to either cost or benefit or both. For a concrete example, the transformation of electrical energy into heat may appear in the constraint models defining a power distribution grid not as a contribution to the penalty figures but as a contribution to the benefit model of a heating system.

The output penalty and merit figures generated by a network's constraint models are where the end-to-end and systemwide attributes of a network first emerge into view as a network is defined. The concomitant feedback or signaling mechanisms may be implicit or explicit and may be any mixture of in-band and out-of-band. In-band feedback mechanisms are those exploiting signaling explicitly carried or implied by components of the resources exchanged.

Note that the existence of constraint models is an inherent factor in network science research and one of the dominant reasons for interdisciplinary approaches. All networks implicitly or explicitly have one critical set of dynamic cost models: The links have finite speeds for exchanging resources and nodes have finite throughput. In particular, the costs derived from these models ensure that the processing of exchanges and the actions of the feedback mechanisms occur locally (see Table D-10).

Driving Dimensions

The analysis of the questionnaire responses also identified three additional significant and common dimensions that are drivers of the difficulty of many associated challenges and of the research effort to address them: complexity, scale range, and network context.

Note that these three dimensions, although widely mentioned in the responses and critical to the challenges and

TABLE D-10 Summary Decomposition of Constraint Models

Constraint Models	Structure	Dynamics
Cost models, benefit models, and the associated models of feedback effects.	Models for determining the costs and benefits for a given node of the locally visible neighborhood.	Models for determining the costs and benefits of adding, deleting, and modifying nodes and links.
	Models for determining aggregate costs and benefits of the network nodes, links, and their properties.	Models for costs and benefits for the exchange of classes of resource, with its associated transformation.
	Models for how these costs and benefits affect network structure and its evolution.	Models for how these costs and benefits affect network behavior and its evolution.

potential value of a possible discipline of network science, are not required for a system to be studied as a network. The greatest benefit from a rigorous network science, however, would lie in understanding the laws that drive the structure and dynamics of networks across the extremes of these three critical dimensions:

- *Complexity.* This dimension includes issues such as large scale (a large number of nodes, links, classes of exchanged resources, or constraints), as well as how the nodes and links behave.
- *Scale range.* This dimension reflects the wide range of interacting critical temporal and spatial scales in the structure and dynamics of a network.
- *Network context.* This dimension addresses the environment of a network as it relates to other networks: Most networks exist in the context of a larger set of other networks on which they depend and with which they interact.

These networks may be naturally captured as different levels of abstraction or as competing and cooperating networks at the same level of abstraction. A social network, for example, may be strongly influenced by the characteristics of the communications, economic, and transportation networks in which the social organisms are embedded (and they, in turn, affect those networks), but each network is best dealt with to a first approximation as its own form of abstraction, using appropriate approximations to reflect how the other networks affect the exchange, storage, and transformation of its various classes of resource of interest (see Table D-11).

Driving Applications

The 427 responses through April 29, 2005, that proposed driving applications for network science (68 percent of all responses, including 92 percent of those who said an identifiable field exists and 10 percent of those who said it does not) described a highly disparate set of applications, generally tightly bound to specific other fields or problem areas (see Figure D-9). Most of the responses were fairly terse and high level and showed little consensus on specific applications. The committee's analysis identified five major communities of research players: technological, biological, social sciences, interdisciplinary, and physical sciences and math. When viewed at a level high enough to allow identifi-

TABLE D-11 Summary Decomposition of the Problem Dimensions of Networks

Dimension	Structure	Dynamics
Complexity	A high number of nodes, links, resource classes, or rules in the cost models and benefit models.	High numbers of internal states and transition rules for the behavior of nodes and links.
Scale range	Dependencies across wide range of spatial dimensions. Highly disparate node and link attributes within one network.	Dependencies across a wide range of timescales. Highly disparate rates of interaction and evolution in different regions of the same network.
Network context	Number and nature of peer networks and of networks at other levels of abstraction. Opacity: unavailability of information or resources across network boundaries.	Definition of behavioral rules governing resource exchange, constraints, and interactions between nodes.

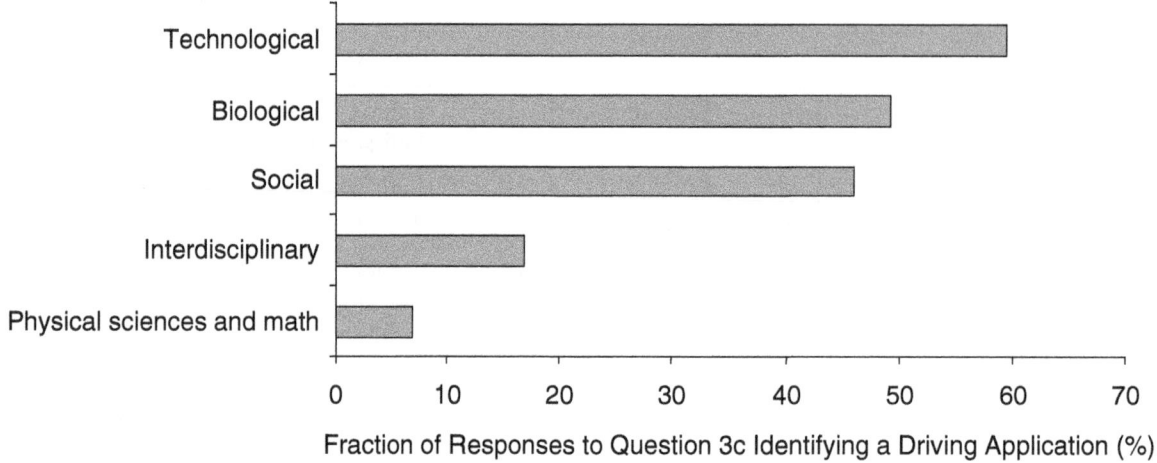

FIGURE D-9 Driving applications identified by respondents.

cation of shared concerns, the current driving applications proved to be closely related to the description of the major research challenges (covered in the next section).

There were also a few voices dissenting on the question itself. For example, one response was that there are hundreds of applications called network science; another was that applications are not the drivers for network science, as it is still an emerging basic research field. In contrast to the view that there were too many applications to consider, another respondent's view was that more effort has been spent on the search for universality principles in networks than on the rigorous study of stand-alone application areas. These dissents echoed the reasons given for saying there is no such field as network science.

Because so few responses were received from outside academia, no useful conclusions can be drawn about the interests of specific nonacademic communities. Within the academic world, the players are generally grouped into well-defined communities focused on particular domains of study. For this reason, the frequency with which particular classes of applications are cited in the questionnaire response closely tracks the response rate by field shown in Table D-4. Within each community, the leading applications in terms of the number of responses are shown in Figure D-10 and Table D-12.

The leading concerns of the technology and engineering communities are closely related. Distributed computing focuses on the efficient realization of applications that are

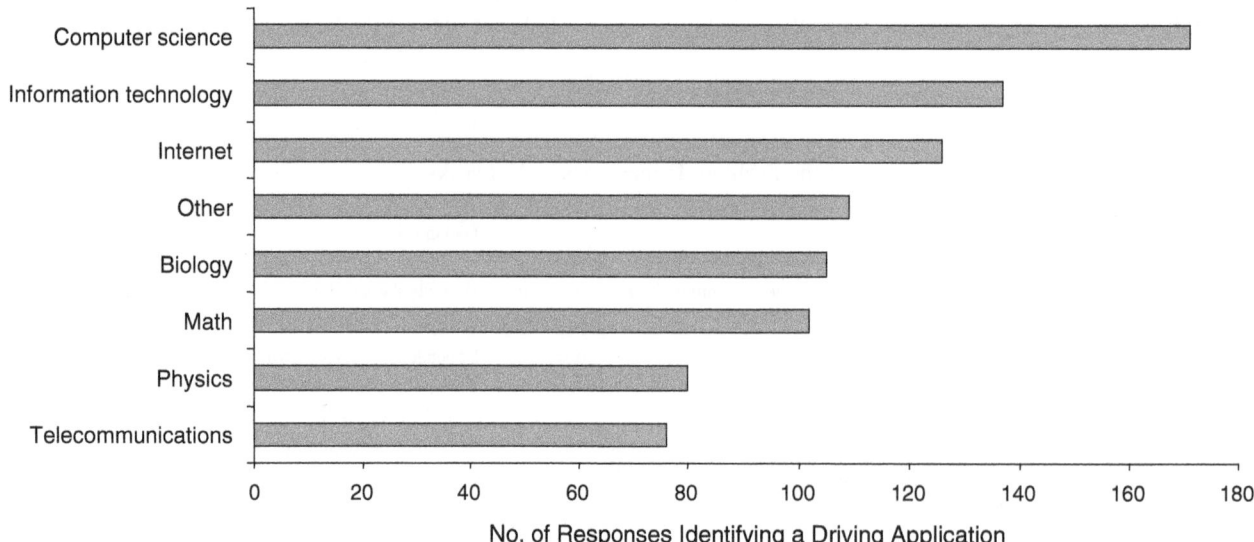

FIGURE D-10 Number of responses to driving applications question.

TABLE D-12 Major Players and Cited Applications

Players	Summary of Most Cited Applications
Technological	Distributed computing Information sharing and discovery Telecommunications
Biological	Public health and disease transmission Ecosystem modeling Systems biology
Social sciences	Social network analysis Economic models and resource distribution
Interdisciplinary	Understanding complex systems Intersection of human interactions and networking technology
Physical sciences and math	High-energy physics Mathematical models of networks

highly decentralized; many of these, in turn, are entwined with issues of networked information sharing, such as peer-to-peer methods. The classic network problems of telecommunications engineering (network design, reliability and resilience, cost-performance trade-offs) appear frequently, but generally with an emphasis on wireless infrastructures (including self-organizing ad hoc and sensor networks). Network collapse (meaning the collapse of the ability to transport the communications payload) is also involved.

The driving applications pursued in the biological sciences include models of disease transmission, ecological modeling and biodiversity, and systems biology. The focus of network understanding of disease transmission is both predictive—How will epidemics spread? What is the relation between the structure of the transmission network and the evolution of the disease?—and interventional—What changes to the underlying social and transportation networks would prevent or reduce epidemics? Ecological applications focus on understanding the flows of energy and nutrients in ecological networks, the interdependence of organisms and species, and qualitative changes in an ecosystem (such as biodiversity stability and collapse). Systems biology refers to the need to understand the system-level architecture of a cell or an organism, as well as to design drugs and interventions that can cause the desired effects and very few side effects. However, a few responses expressed concern about the time required to go from basic system biology to specific medical applications.

The applications in the social sciences are affected by the deep history of the discipline's analysis of social interaction and influence networks. Other respondents were concerned with understanding the well-known, heavy-tailed distributions in numerous social constructs in terms of the underlying social networks. Economic network issues such as the flow of capital also relate to the comparative impact of underlying infrastructure networks for communications, application information sharing, and transportation of people and material. Network science is being applied to distribution channel behavior, such as interpersonal ties within a market or interorganizational ties in a value chain. Smaller but still significant numbers of responses mentioned organization models and political applications, ranging from disrupting terrorist networks to supporting prodemocracy organizations under authoritarian regimes.

"Interdisciplinary communities" refers specifically to respondents who self-identified as being involved in several disparate fields and/or who proposed application topics that explicitly span or relate disparate fields. This distinction is necessary because many responses gave lists of unrelated applications drawn from different fields. In this category the committee included "complex systems," which occurred commonly but was generally not given further definition; the term "emergent phenomena" is closely related and likewise undefined by the respondents. Descriptions of driving application also commonly mentioned the need to understand the relationship between disciplines such as biological and computer networks.

The use of telecommunications networks for data-intensive computing applications was cited as one driving application on the physical sciences; other such applications in that area spanned physics and chemistry, such as understanding high-dimensional dynamical systems and the network structures underlying the energy landscapes that drive protein folding and similar optimization behaviors. One driving application in mathematics was theories for relating system-wide behavior and network structures.

RESEARCH CHALLENGES

The committee, aided by the questionnaire, identified a number of important research challenges that should be addressed if the field of network science is to be moved forward. The analysis of research challenges was based on the responses to question 3d (What are the key research challenges?) and was performed by reading through all of the responses to the questionnaire and binning the results to infer broad topics that recurred frequently. To ensure that the responses were not being biased by individual committee members, the responses were compared with an earlier, independent analysis of the responses.

The responses that fit within a broad category of challenge were counted, and the seven most highly populated categories were selected for inclusion in the report. Each category had between 25 and 100 responses identifying it as a research challenge; the proportion of responses garnered by each challenge is shown in Figure D-11.

The seven primary challenges that were identified were these:

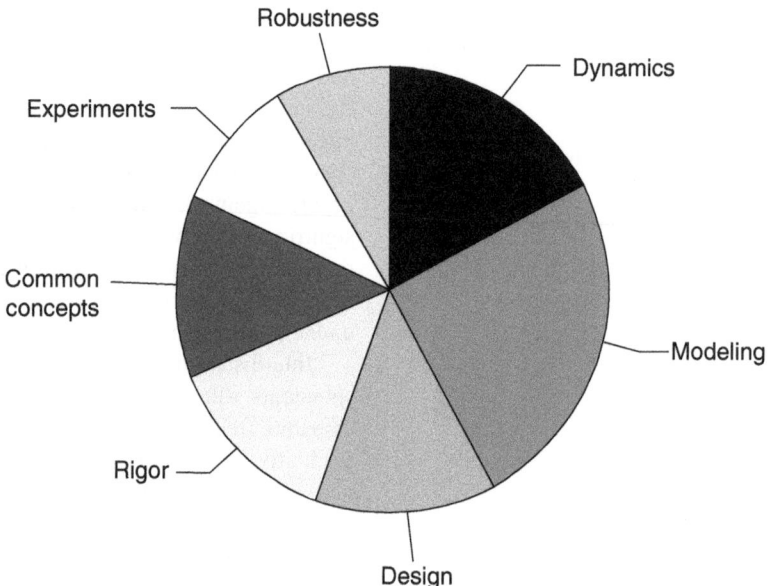

FIGURE D-11 Major research challenges.

- *Dynamics, spatial location, and information propagation in networks.* A major need in network science is a better understanding of the relationship between the architecture of the network and its function. This is particularly important in networks where dynamics plays a large role, either through the flow of information around the network or through changes in the network structure (by evolution or adaptation). How the structure of a network relates to the behavior of the system is still not well understood and will be a major impediment to progress in many applications.
- *Modeling and analysis of very large networks.* Present tools and approaches are designed to work with relatively small networks, but many of the most important problems involve much larger networks. Examples of such networks include cell regulatory networks in biology, social and economic networks, and computer communication networks (including military command-and-control networks). Abstractions and approximations are needed that allow reasoning about these large-scale networks, as well as techniques for modeling networks with noisy and incomplete data. Analysis techniques for such networks must have good scaling properties so that they can be applied to the very large networks that are key to network science.
- *Design and synthesis of networks.* While a lot of engineers, scientists, mathematicians, and sociologists are simply trying to understand complex networks already in existence, in many application areas the goal is to design a network to obtain a desired behavior—for example, scalability, robustness, usability, resiliency, efficiency, and resolvability (or adaptability). It may be possible here to learn from biological systems how to design engineered systems that exhibit equally complex, adaptive, and robust behavior.
- *Increasing the level of rigor and mathematical structure.* Many of the respondents to the questionnaire felt that the state of the art in network science did not have an appropriate mathematical basis. This level of mathematical rigor could be achieved by a combination of defining the appropriate levels of abstraction for analysis, developing better tools in graph theory and other relevant disciplines, and searching for fundamental limits of performance.
- *Abstracting common concepts across fields.* Many members of the committee and respondents to the questionnaire cited the need to define common concepts across the disparate disciplines and applications that are part of network science. The multidisciplinary nature of the work is a challenge, but results could be transferable from one field to another if appropriate unifying principles can be developed.
- *Better experiments and measurements of network structure.* Current data sets on large-scale networks tend to be sparse, and tools for investigating the structure and function of these networks are limited in many domains. There was a general feeling shared across many fields that there needs to be more and better access to data, which in some domains requires new measurement techniques to be developed—for example, to obtain a detailed spatiotemporal measurement of the operation of a cell. One respondent suggested the de-

velopment of a so-called "macroscope" to detect, communicate, and understand the structure and dynamics of large-scale networks.
- *Robustness and security of networks.* Finally, there is a clear need to better understand and design networked systems that are both robust to variations in the components (including localized failures) and secure against malicious intent. This requires a much more sophisticated understanding of the failure mechanisms in networked systems as well as better tools for predicting the impact of perturbations on networked systems.

The Social Structure of Network Science

In addition to the analysis performed by the committee, the data were also analyzed by Katy Börner, who addressed the visible social structure of research in network science as indicated by the collaboration and invitation entries of each respondent. Her analysis is provided in Box D-2. Upon reviewing her analysis, the committee consensus was expanded to include the following two findings presented in Chapter 6 on the empirical state of the proposed field of network science:

Finding 6-7. Analysis of the social and collaboration networks of the respondents provides additional evidence that network science is an emerging area of investigation.

While the clusters within the network are only weakly connected, a large connected core spans many of them. Based on Dr. Börner's extensive experience, and on the judgement of the committee, this pattern is characteristic of an emerging field and constitutes objective evidence that network science is a field, but an immature one whose future is still undecided.

Finding 6-8. Analysis of the social and collaboration networks of the respondents provides additional evidence of the multidisciplinary nature of network science.

Dr. Börner's analysis of the social and collaboration networks provides additional evidence of the multidisciplinary nature of network science. Researchers from any given discipline are distributed throughout the graph, and any given subcommunity includes researchers from multiple disciplines. This pattern is unlike any field previously analyzed by Dr. Börner. In the committee's judgment, therefore, the pattern constitutes objective evidence that network science is a field that is distinctly interdisciplinary, with research concerns that support multiple application domains.

BOX D-2
Mapping the Social Network and Expertise of "Network Science" Researchers

This box presents the anonymized results of a bibliometric analysis[1,2] of the social networks and expertise coverage of network science researchers prepared at the committee's request by K. Börner and W. Ke, of the InfoVis Laboratory at Indiana University. All results are based on the self-reported data in the file named "cleaned_survey_as_of_050318_0910a_posted.xls." Subsequently, the authors report the data® cleaning and analyses performed, major results, and their interpretation. They conclude with a set of recommended topics for further study.

Data Set Used, Analysis Results, and Interpretation

The data file "cleaned_survey_as_of_050318_0910a_posted.xls" comprises 499 completed questionnaires that report 923 "collab_with" links reported under Q2c and 376 "invite" links reported under Q4a. To ensure a high quality of automatic data extraction and analysis, all names reported in free-form text as "Other collaborators" under Q2c and all "Other people to invite" reported under Q4a were not considered. Figure D-2-1 illustrates relationships among the initial invitees, respondents, and identified collaborants.

In total, 1,241 unique names of network science researchers were identified. E-mail addresses were used to ensure that these names are truly unique and represent exactly one person. As requested by the National Research Council, author names were replaced by a unique identification number to preserve the anonymity of authors.

In addition, the 22 (checkable) fields of interest as well as the free-form text of "other" fields of interest reported in Q1c were analyzed. In total, 138 unique fields of interest were identified. Fields that were mentioned most often were computer science (mentioned 201 times), information technology (166), and Internet (156).

Data Quality Issues The "collab_with" links are mostly made to researchers in spatial or thematic proximity. Hence, these links help to grow the social network of network science researchers locally. Colleagues reported that they tried to "invite" people who were not yet in the data set. There was no question that asked users to identify "major players" or "gatekeepers." There are many misspellings of names and disciplines in this data set. Information provided in the "other collaborators" and "other people to invite" section could not be used in this automatic analysis.

Data Analysis Results Here we report "major researchers" who are frequently mentioned in the data set, who act as gatekeepers, and who interlink many scientific fields. In addition, we extracted and will present existing social and collaboration networks. Researchers who are frequently mentioned in the complete data set and the number of times they are listed as a collaborator are given in Table D-2-1. Figure D-2-2 shows the major components (size≥10) network of the network science researcher network (NSRN). The Pajek[3] plot shows exactly 630 of the 1,241 unique researchers, and their "collab_with" links and "invite" links are shown. Each researcher is represented by a node. Node color coding is used to identify researchers that submitted (brown) or did not submit (orange) questionnaires. The node areas' size corresponds to the number of times a researcher is mentioned by other researchers. Each

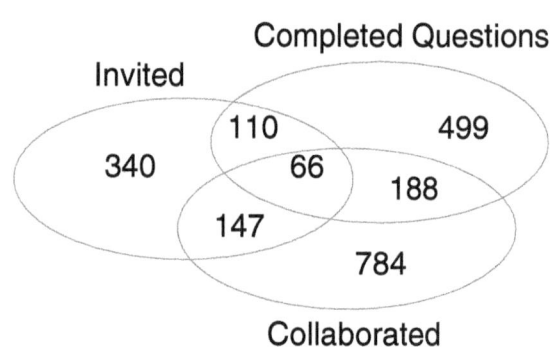

FIGURE D-2-1 Relationships among invitees, respondents, and collaborators.

TABLE D-2-1 Researchers Who Are Frequently Mentioned and Listed as Collaborators

ID	No. Listed	No. Listed as Collaborator
1005	12	8
9	8	7
512	12	6
1009	7	5
139	7	5
1023	8	5
1047	5	4
784	6	4
455	6	4
814	4	4
1238	7	4
925	5	4

BOX D-2 Continued

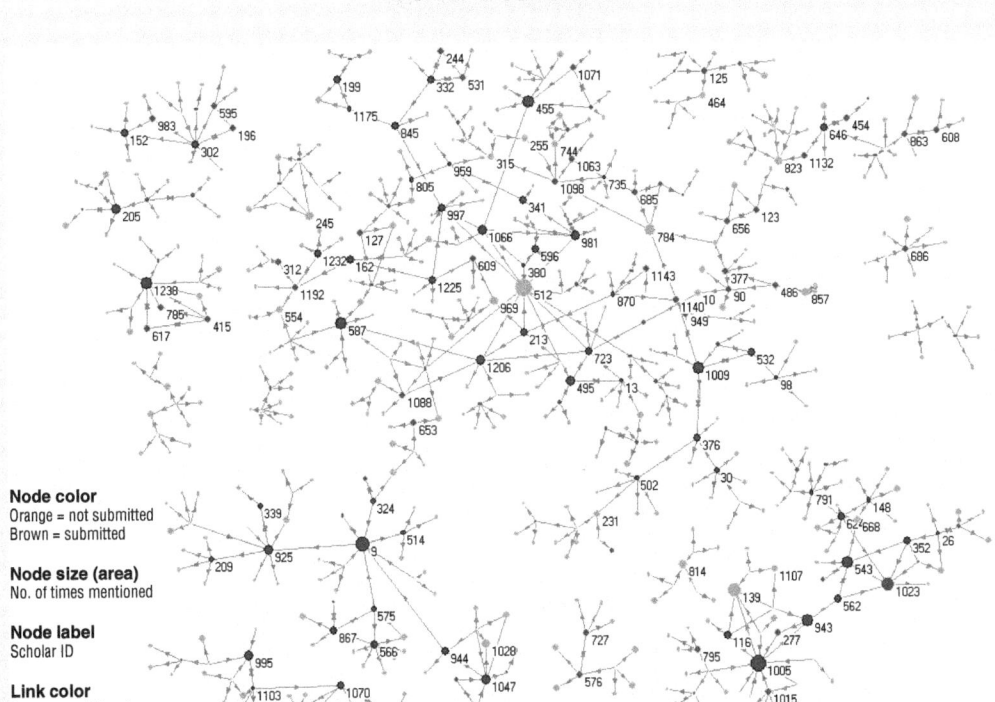

FIGURE D-2-2 Network science researchers network.

node with a betweenness centrality no less than 0.00001 or a size (number of appearances in the data set) ≥3 was labeled with the author's name (ID). Links denote "collab_with" links (in orange) and "invite" links (green).

Subsequently, researchers who act as gatekeepers were identified based on an examination of the betweenness centrality (BC) values[4,5] of nodes in the NSRN. The top 10 researchers are given in Table D-2-2. Figure D-2-3 indicates nodes with a BC value ≥0.00001 by a black ring and shows them in the context of the NSRN.

To examine the community structure of network science researchers, we examined the different components in the NSRN. Table D-2-3 shows the size of existing components, the number of components that have this size, and the total number of nodes in these components. The largest component in the NSRN is shown in Figure D-2-4 using the color coding introduced in Figure D-2-2. It represents the current coherent core of the new field of network science.

TABLE D-2-2 Researchers Who Act as Gatekeepers

ID	No. Mentioned	Betweeness Centrality Value
1066	4	0.00020275
997	2	0.00017878
981	4	0.00015093
9	7	0.00013408
341	2	0.00012502
925	4	0.00010882
845	3	0.00010688
959	1	0.00009716
1225	2	0.00007773
162	3	0.00007060

box continues

BOX D-2 Continued

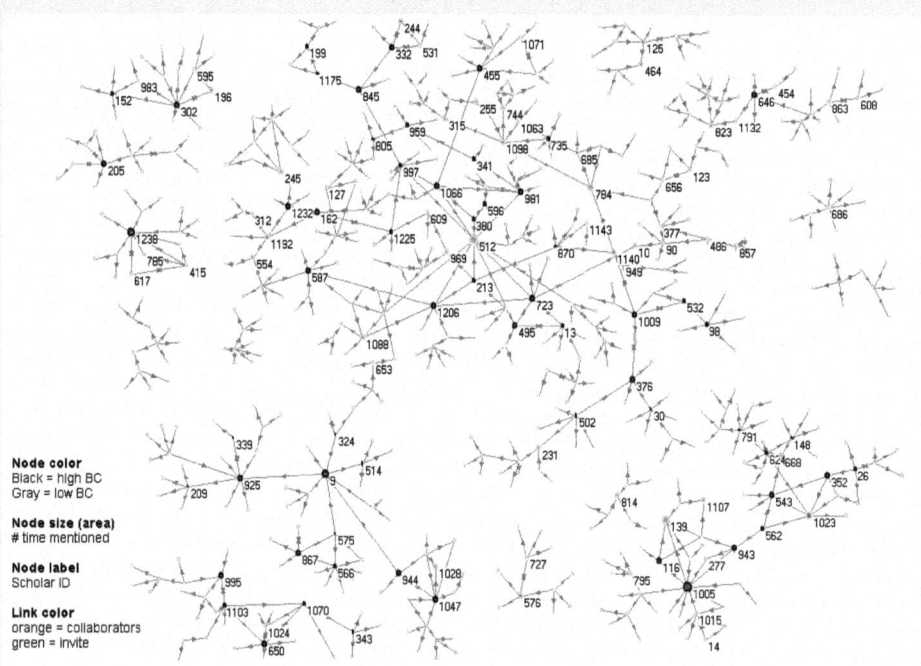

FIGURE D-2-3 Researchers with high BC values (in black) and low BC values (in gray).

TABLE D-2-3 Components in the NSRN

Size	No. of Components	No. of Nodes
1	77	77
2	32	64
3	25	75
4	45	180
5	10	50
6	12	72
7	7	49
8	1	8
9	4	36
10	4	40
11	2	22
13	1	13
14	1	14
15	1	15
17	1	17
18	1	18
30	1	30
33	1	33
73	1	73
355	1	355
Total		1,241

box continues

APPENDIX D

BOX D-2 Continued

Interpretation Compared with maps of other scientific disciplines, the NSRN clearly exhibits the characteristics of a new and emergent research area: It consists of many unconnected networks of collaborating network science researchers, and the existing networks show a rather heterogeneous coverage of research topics.

Figure D-2-5 is a map of all network science researchers visualized in VxInsight.[6] The map at the left-hand side shows the NSRN. On the right, the very same graph is shown in "landscape" mode, with colored dots representing the self-reported interest profiles of researchers. A white dot denotes that the researcher listed "biology" as a principal field of interest in Q1c. Yellow denotes "computer sciences," light blue "Internet," blue "physics," and green "sociology." As can be seen, there are no groupings of researchers with similar fields of interest. Instead, very different research interests seem to be almost equally distributed over the NSRN.

As the field of network science matures, subareas devoted to the study of specific research fields are likely to emerge, and many of the separate components will exhibit collaboration links, weak or strong and temporary or stable.

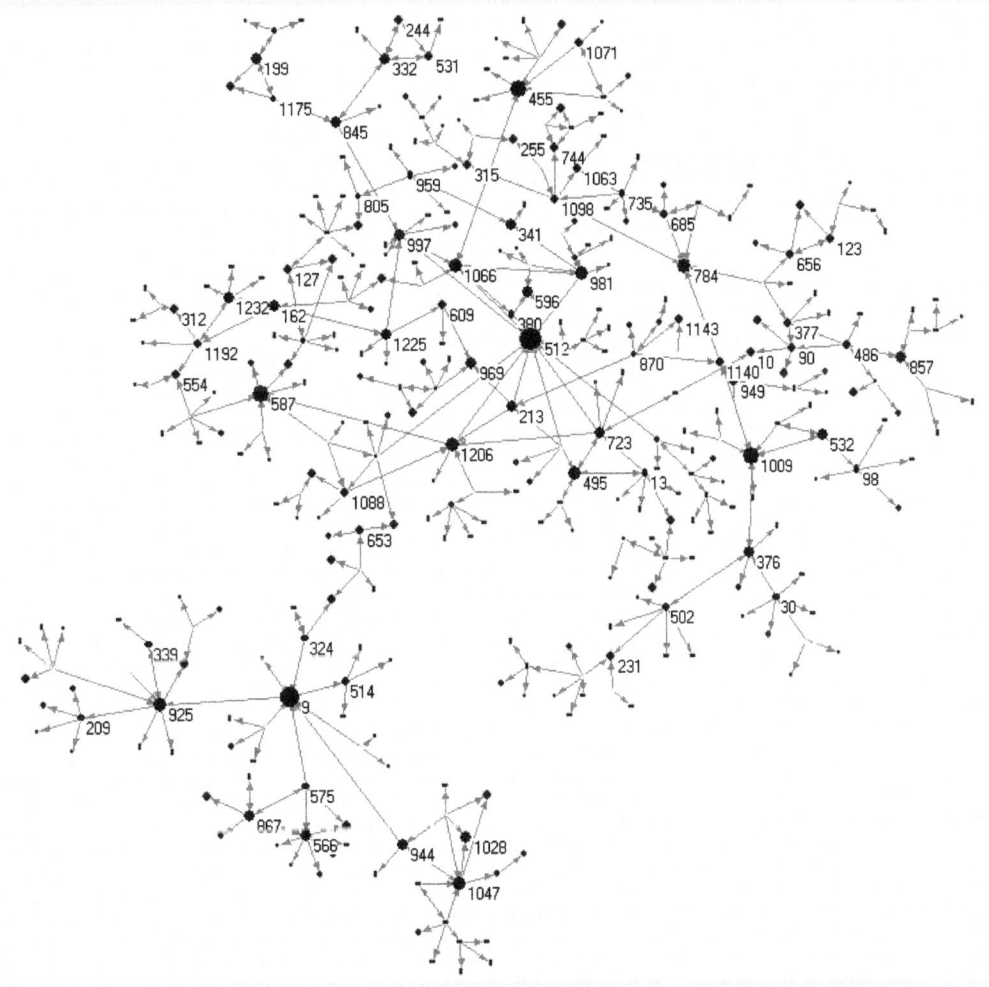

FIGURE D-2-4 Largest component of the NSRN.

box continues

BOX D-2 Continued

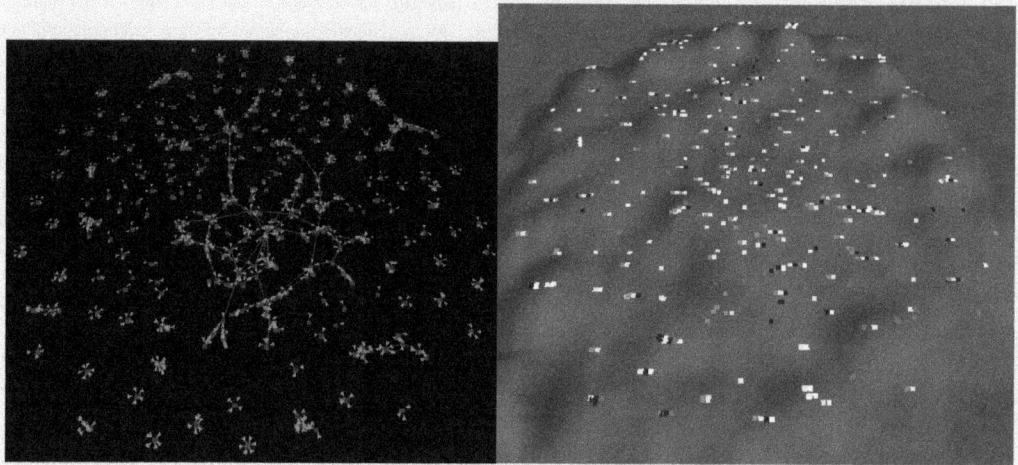

FIGURE D-2-5 Disciplinary heterogeneity of the NSRN.

Social Network of Network Science Researchers: Topics for Further Study

• *Increase our understanding of the interplay of affiliation, thematic, and social interrelations among today's network science researchers.* Invite key network science researchers to identify and label the main research groups key shown in Figure D-2-2.

• *Bibliometric analysis of networkscience publications, patents, and funding data.* The InfoVis Laboratory at Indiana University is developing the sociotechnical infrastructure to analyze the structure and evolution of scientific disciplines and all of science on a large scale.[7] Major publication, patent, and grant databases (covering mostly U.S. research) are available, as are scalable algorithms and compute power. A detailed, objective analysis of scholarly data would complement the self-reported, subjective data and its analysis reported here.

• *Development of an online portal that tracks and communicates the evolution of network science research and results.* Geospatial and semantic maps of network science researchers and publications presented here and proposed in Shiffrin and Börner (2004) can be made available online as a unique interface to data sets, publications, and expertise related to network science research. Researchers interested in being "on the map" should be given the option to submit data about their publications, collaborators, etc. The incentives for researchers to contribute high-quality data can be further increased by using this online map to make funding decisions much as PI's resumes are used today. Assuming that a comprehensive set of high-quality data can be acquired on a continuous basis, an interactive, continuously evolving, weather-forecast-like map of network science research can be made available to grant agencies, researchers, practitioners, and society at large.

NOTE: The authors would like to thank Will E. Leland for compiling the data set used in this study and for insightful feedback on previous results. This work is supported by a National Science Foundation CAREER Grant under IIS-0238261.

[1]Börner, K., C. Chen, and K. Boyack. 2003. Visualizing knowledge domains. In Annual Review of Information Science & Technology, B. Cronin, ed. Medford, N.J.: Information Today, Inc./American Society for Information Science and Technology.

[2]Shiffrin, R.M., and K. Börner, eds. 2004. Mapping knowledge domains. Proceedings of the National Academy of Sciences of the United States 101 (Suppl. 1).

[3]Batagelj, V., and A. Mrvar. 1997. Pajek: Program package for large network analysis. Available at http://vlado.fmf.uni-lj.si/pub/networks/pajek/.

[4]Freeman, L.C. 1997. A set of measuring centrality based on betweenness. Sociometry 40: 35–41.

[5]Brandes, U. 2001. A faster algorithm for betweenness centrality. Journal of Mathematical Sociology 25(1): 163–177.

[6]Davidson, G.S., B. Hendrickson, D.K. Johnson, C.E. Meyers, and B.N. Wylie. 1998. Knowledge mining with VxInsight: Discovery through interaction. Journal of Intelligent Information Systems 11(3): 259–285.

[7]Available at http://iv.slis.indiana.edu.

E

Opportunities for Creating Value from Network Science

This appendix supports the findings of Chapter 5 and provides details of the scenarios developed in response to Items 3 and 4 in the statement of task. What are the viable investment options for the Army and nation in network science? How does exploiting these opportunities create value? The committee attempted to answer these questions by constructing three scenarios that represent fundamentally different levels of response by the Army.

Scenario 1, Building the Base, examines the alternatives for the Army if it wants to invest a modest amount (on the order of $10 million annually) in 6.1 basic research within the context of its present funding processes. While some changes in the style and content of its present research activities are envisaged, this scenario represents a minimal investment and a minimal change in posture relative to the current Army R&D portfolio of investments and processes. It creates value by increasing the knowledge of topics in the core content of network science and by expanding the base of trained practitioners.

Scenario 2, Next-Generation R&D, examines alternatives created by applying best practices in modern industrial R&D to the Army's R&D investments in network science. In this scenario, basic research (6.1) investments are coupled with larger applied research (6.2) and, possibly, 6.3 investments, to create technology investment options for the Army to exercise in concert with suppliers. The committee envisages, however, that the management of this process would be substantially different from present Army practices. The committee gives an idea of how these changes could be effected by creating a market-driven R&D environment. Projects that might be funded in this environment are exemplified by three kinds of projects: one in social networks, one in engineered networks, and one in biological networks. Comparison of these projects shows the sorts of issues and benefits that could arise from Scenario 2 investments in the three different areas of important network applications.

Scenario 3 takes as its point of departure the phrase "to enable progress toward achieving Network Centric Warfare capabilities" in Item 4 of the statement of task. In it, the committee asks what the Army (or, more appropriately, the DOD Office of Force Transformation [OFT][1]) should do if its primary objective is to create these capabilities in a robust and affordable way over the course of the next decade. A significant investment in network science would be one element in such an effort, but the emphasis would reside elsewhere. Specifically, in this scenario the committee examines past efforts of the federal government to develop major new warfare capabilities all the way from basic research to deployment in the field in order to estimate the magnitude of the effort and investment required to source and deploy such capabilities.

SCENARIO 1, BUILDING THE BASE

Scenario 1 involves a modest level of funding (~$10 million per year) that fits into the Army's current scheme for 6.1 basic research. Small amounts of Army risk capital funds are invested to create a knowledge and personnel base from which it can attack the practical problems that arise when trying to provide network-centric warfare (NCW) or network-centric operations (NCO) capabilities. It is for this reason the scenario is called "building the base."

Because the anticipated investment is too small to fund significant interdisciplinary efforts, it should be focused on leveraging existing research in areas related to network science. It would support 20 to 30 single-investigator or small-group grants of $300,000 to $500,000 per year. Such an investment would allow the Army to tap into top research efforts in physical, social, and biological networks. The Army should fund only the most outstanding researchers because the very best are qualitatively different from even the very good. The essential point is that to pay off big in

---

[1] For further information, see http://www.oft.osd.mil. Accessed August 19, 2005.

terms of breakthrough ideas and models, the very best researchers must be recruited to do the work. To attract an elite group who will influence the future of network science, the Army must adapt its funding and management policies, possibly offering larger and longer-duration grants with few or no restrictions on citizenship, publication, and the like. The selection criteria would be excellence of research as judged by peer review and track records, relevance to Army objectives, and opportunities for Army involvement through inclusion of suitable Army personnel in the effort.

The committee envisions that the research efforts would be located at major research universities. An important aspect of the program might be a once-a-year conference at an Army laboratory or facility, where the principal investigators (PIs) would report on accomplishments during the year. The program needs enlightened management to support interdisciplinary work through the interaction of a diversity of PIs. A major theme of the program would be the achievement of fundamental advances in network-research-based statistical physics, applied mathematics, development of mathematical models of social phenomena, and other areas by generously funding only exceptionally talented individuals who are organized into a national social network.

Such a program would be the first to address the needs of network science per se. The program would focus on networks as coherent entities characterized by their architecture, structure, and dynamics. By deliberately adopting a broad theoretical and methodological focus, the program would encourage the creation of fundamentally novel ideas. A wide diversity of approaches is a key feature of long-term success. Keeping the goals broad and flexible would allow the Army to cultivate such diversity; a narrowly defined program would eliminate much of the creative potential that could lead to breakthroughs and new ideas. One way to do this would be for the Army to hold one or more workshops with appropriate personnel to determine the topical content for such a program.

The Army's needs are broad and fundamental in nature: It must learn how to approach the creation of a predictive description of large, interacting, layered networks. A basic science program is the first step toward building up the critical mass of talent needed to address specific Army problems in this area. This modest approach would allow the Army to identify a relevant research community, and organize it so that, with time, it could be called upon to address specific needs.

The proposed approach differs from existing programs in agencies such as the National Science Foundation (NSF) and the National Institutes of Health (NIH) in that it focuses on network science per se. While a significant amount of research is taking place in communities addressing the applications of networks, almost none of this research is funded by dedicated network science programs. Instead, researchers are funded by, for example, the NSF, to improve the Internet or to understand the statistical mechanics of complex systems, or the NIH, to uncover the features of specific organisms or biological processes. Because many of these research programs cannot avoid dealing with the network aspects of these problems, they divert some of their funding in that direction. Nevertheless, most of their work is focused on applying network ideas to specific systems rather than on developing new tools and ideas. Currently, no agency the committee knows of has a specific program devoted to network science. A research program on network science per se is a new concept, with unique and novel goals.

As a consequence of its discussions with Army and DOD representatives, the committee has come to realize that the fundamental problems underlying effective NCO lie in the social domain. Yet how people interact and utilize technology or make decisions based on shared knowledge are areas almost unexplored in the Army's current basic research portfolio. The committee urges the Army to focus additional resources on the possible applications of network thinking to social, especially organizational, issues (Helper et al., 2000).[2] Applications to biology, engineering, and the physical sciences are also essential to Army applications, but the Army is already funding research in these areas. The committee suggests that on the margin the most significant problem is not how to build better satellites, tanks, or medicines but rather is how to organize millions of individuals to collect intelligence, deliver supplies, and prosecute wars over an increasingly global, constantly shifting geographical and political playing field (Garstka and Alberts, 2004). This is a monumental problem that has not, however, traditionally been the province of science. Rather, it has been managed through a mixture of intuition, experience, and tradition. A significant fraction of the proposed program should address this organizational problem the way scientific problems are addressed: through a combination of theoretical modeling, data analysis, and controlled experimentation.

To illustrate the flavor of its thinking, the committee closes this scenario by indicating some promising research topics in four broad areas: network structure, network dynamics, network robustness and vulnerability, and network services. Each area has theoretical, empirical, and experimental components. Investments in basic research in each can provide value for the Army.

Network Structure

Advances in the applications of network research are limited by a lack of fundamental understanding of the structure, evolution, and topology of complex networks (Newman, 2003; Watts, 2004; Newman et al., 2005). Many network models have been formulated and studied by numerical

---

[2]C.F. Sabel, professor of law and social science, Columbia Law School, "Theory of a real-time revolution," briefing to the 19th EGOS Colloquium, Copenhagen, July 2003.

simulations or approximate analytical methods (Albert and Barabási, 2002; Pastor-Satorras and Vespignani, 2004; Dorogovtsev and Mendes, 2003; Ben-Naim et al., 2004; Bornholdt and Schuster, 2003; Barabási, 2003). While these efforts constitute an advancement in the modeling and representation of networks, the topology of these models is far less rigorously understood. Questions concerning the universality of topological properties, correlations introduced by dynamical processes, and the interplay between clustering, hierarchies, and centrality in networks remain only partially answered (Amaral et al., 2004). A full understanding requires the development of rigorous methods to uncover the mathematical structure of growing networks. Advances in this direction can be achieved only by the empirical study of real networks and by developing the appropriate mathematical tools and metrics not only to characterize novel systems but also to classify networks based on their topology, structure, function, and dynamics. In this respect we are lacking in the fundamental characterization of network topology, in our ability to characterize weighted networks and communities in networks, and in our description of the time evolution of the network topology. In parallel, we need to develop discrete and continuum models that generate networks whose structure and evolution mimic those of real networks.

In spite of considerable effort in social network analysis over the past 50 years and great interest in the past decade, surprisingly little is known about either the detailed structure or the time evolution of social networks. In part, the gap in knowledge has come about because purely social interactions, unlike interactions conducted on an engineered network such as the Web, are much harder to track empirically. It also derives from a tendency to view all networks as static entities with structural features rather than as dynamic processes with parameterized rules. Finally, it has resulted from the tendency to ignore the role of social and organizational contexts in network formation and to focus instead on the role of exceptional individuals or small-group structures. Without a better understanding of social networks, it is extremely difficult to advance our understanding of any social processes, including implementation of the cognitive and social domains of NCO to improve combat effectiveness.

Network Dynamics

Networks are specified not only by their topology but also by the dynamics of information or traffic flow along their links. The amount of traffic characterizes the connections in a communication system or large transport infrastructure and is fundamental for a complete description of these networks (Barrat et al., 2004; Almaas et al., 2004). Heterogeneity in the intensity of connections may also be important in understanding social systems. The ultimate objective is a mathematical characterization that might uncover general principles describing the dynamics of networks. Furthermore, networks provide the substrate on top of which the dynamical behavior of the system must unfold. At the same time, however, the various dynamical processes are expected to affect the network's evolution. Dynamics, traffic, and the underlying topology are mutually correlated. It is vital to define appropriate quantities and measures capable of capturing how all these ingredients participate in the formation of complex networks. To carry out this task, we need to develop large empirical data sets that simultaneously capture the topology of the network and the time-resolved dynamics taking place on it.

Understanding "how things spread" through different kinds of social, organizational, and technological structures, and with what consequences, is of paramount concern in an increasingly global economy and culture (Watts, 2003). Examples include electronic viruses spreading through computer or phone networks, biological viruses (weaponized or natural) spreading through the contact network in a society, and ideas or strategic concepts spreading on a social network (Arquilla and Ronfeldt, 2001; Pastor-Satorras and Vespignani, 2004; Anderson and May, 1992). Uncovering the similarities and the differences between electronic virus spreading, biological and social contagion, and the interplay between the network and spreading processes is crucial for the research community. Such research has important potential Army applications, from defending the communication infrastructure against viruses to developing scenarios for limiting the impact of a biological contagion.

Individuals making strategic decisions in uncertain or ambiguous environments (which is to say, almost all environments) are influenced either explicitly or implicitly by the decisions of others (Katz and Lazarsfeld, 1955; Watts, 2003). And when all individuals are paying attention to one another and constantly making or updating various decisions accordingly, "social contagion"—the spread of ideas, influence, norms, innovations, attitudes, rumors, unrest, violence, extremism, etc.—becomes not only possible, but pervasive (Rogers, 1995; Gladwell, 2000; Huckfeldt et al., 2004). In some ways, social contagion seems like biological contagion, but in other ways it is fundamentally distinct (Dodds and Watts, 2004, 2005). Sorting out these differences is a key to understanding network contagion.

In order to exploit the information infrastructure capabilities envisaged for NCO, the Army must comprehend how large groups of individuals coordinate their efforts to solve large-scale problems, where "large-scale" implies that the scope of the problem is beyond the capacity of any particular individual to solve or even fully comprehend (Dodds et al., 2003; Watts, 2003).[3] Examples of such problems include (1) collating large volumes of ambiguous information, collected in widely dispersed environments, to assess the seriousness

---
[3]C.F. Sabel, professor of law and social science, Columbia Law School, "Theory of a real-time revolution," briefing to the 19th EGOS Colloquium, Copenhagen, July 2003.

and immediacy of specific threats; (2) coordinating large numbers of troops in novel and rapidly evolving combat environments; (3) coordinating geographically and industrially diverse supply chains to deliver highly specific and time-sensitive logistical support that varies rapidly and unpredictably; (4) business firms attempting radical innovations or recovering from catastrophic supply-chain breakdowns; and (5) educational or community development organizations attempting change to alleviate widespread and deeply rooted social or economic disparities. Large-scale problem solving is complicated because it involves the study of collective behavior that is oriented toward particular (if imprecisely specified) outcomes rather than just the study of processes in action. It is also, and perhaps not surprisingly, the problem area about which least is understood theoretically.

Biology also deals with diverse large-scale networks. The most prominent cellular networks are metabolic networks, the regulatory network, and signaling and protein interaction networks (Barabási and Oltvai, 2004). The key question here is, What do network science results tell you in terms of biology? Are they falsifiable? Do they offer testable predications? In order to be of value, the predictions must ultimately be testable in the laboratory. This kind of work is being partially funded by the NIH roadmap initiatives. Results in this domain might lead to new ways to analyze networks and to make predictions relevant to rapid response to disease and biological warfare.

Another important structural and dynamical issue is the location of function within a biological network—that is, What is the architecture of the network? Networks typically comprise different groups of interconnected elements, or modules, each one being responsible for different functions. In engineered networks, this architecture is designed in a priori. In biological networks, it must be discovered by, for example, identifying highly interlinked communities of elements. Modules can be repeated at different hierarchical levels and interconnected via the hubs of the system. How modularity emerges across many biological networks and how it can be reconciled with the other properties of these networks are basic questions. The constraint and facilitation of biological phenomena by network structure is an issue that is likely to be at the forefront of biology in the coming years.

Network Robustness and Vulnerability

Networks need to be able to function in spite of errors, failures, and attacks. Such phenomena can affect a network's ability to function in several fashions—for example, by incapacitating nodes and links or limiting traffic, leading to local and potentially cascading failures (Watts, 2002; Dodds et al., 2003; Albert et al., 2000; Cohen et al., 2000, 2001). How do we understand the different modes of network failure? Can we design networks with maximum tolerance of errors and a high survivability under attack? What is the relationship between a network's robustness and its structure? How do the different dynamical processes taking place on networks affect their survivability? These are questions that, if answered, could fundamentally alter the Army's ability to design and implement fault-tolerant communications and organizational networks.

Network Services

The operation of computer networks is governed by an architecture, known as a reference model, in which the functioning of the network is divided into independent layers built on top of each other (Tanenbaum, 2003). The lower layers of this architecture deal with the physical connections (including wireless) and electrical signals transmitted across these connections that link the nodes of the network. The important concept underlying this organizational construct is that each layer of the architecture delivers "services" to the layer above according to a standard interface convention. This organization is important to hide the workings of one layer of the architecture from those of all the others. Thus, the technology used to implement one layer can be upgraded without changing the operation of the layers above and below. In the open system interconnection (OSI) reference model describing the architecture of computer networks, the lowest layer is referred to as the physical layer, and the next two layers as the data link and network layers, respectively. On top of these sits the transport layer, whose function is to ensure that the messages from the layers above get transmitted reliably from one computer to another even if the lower three layers are unreliable. These four layers, in concert, establish a reliable communications link between two computers (nodes) on the network, over which messages associated with services created in the upper layers are delivered (Tanenbaum, 2003).

The Internet is the most important and familiar computer network. As important as the Internet itself, however, are the services that are created at the upper levels of the Internet architecture. Among the most visible services that have evolved on the Internet are e-mail, the World Wide Web (www), search engines, and the targeting and formatting of information to individuals. Other services, such as making information available in digital format and positioning data globally, have added value to the network.

Search engines such as Google play a major role in finding Web sites for travel, medical information, research papers, indeed almost anything you can conceive. As commercial activity on the Web increases, all kinds of services that support it have developed. These include services that provide individualized news, weather, driving directions, and maps. Such services allow one to track a storm as it travels across a state in real time or to view the road surface of a highway pass from a camera mounted at the summit. They are changing the way citizens access and use information. Underlying these activities is an evolving research base in filtering, clustering, ranking, and optimization. This is a broad area of study carried out largely in computer science, electrical engineering, and computer engineering departments worldwide and in many firms. Some of the important topics encompassed by this area are data mining, distributed computing, network protocols, information capture and access, and security, privacy and cryptography. Since these services are vital to NCO, support of basic research in this area should ultimately create value for the Army.

SCENARIO 2, NEXT-GENERATION R&D

Scenario 2 envisages applying best practices in industrial R&D management to the Army's investments in projects that combine basic and applied network science. Specifically, the committee expects the objective of these projects to be the articulation of technology investment options that could be exercised by the Army and its vendors to provide a desired capability. The amount of this investment is envisaged to be between $25 million and $100 million annually, roughly $25 million per project. There are expected to be investments in the university community for the basic research and in both Army in-house activities and commercial firms for the applied research. The committee envisages, however, that the R&D projects would be managed in a way profoundly different from the way in which current Army in-house and external centers are managed.

The selection of projects to be funded would be market driven and controlled by a top-level Army team. It is expected that connections between the basic and applied portions of the research will be much more intimate. Modern Internet collaborative tools would be used to manage the day-to-day work in rough analogy to the global design of industrial products. The activities are managed in small, intimate groups devoted to specific subprojects that are integrated into the overall project in a looser networked fashion. People flow from one small group to another over time. The entire team makes up a social network consisting of smaller, more tightly coupled social networks. In short, this scenario envisages the application of modern communications networks and tools and the insights of modern social network theory to transform the management of Army R&D projects.

The committee proceeded by first agreeing on the most effective management approach for next-generation R&D projects. It developed the concept of "market-driven" management to apply to all projects that characterize the entire Scenario 2 segment of the opportunity space. While associated issues and benefits will vary depending on the network application, the market-driven approach is the same. To illustrate the scope and scale of next-generation R&D with market-driven management, committee members developed three sample projects in the sociological, engineering, and biological areas of network applications.

These projects were selected in diverse areas to underline the committee's belief that research will be equally necessary in all areas to advance network science. While each of the sample projects has the potential to advance network science, they should not be construed as a "shopping list," and the committee does not recommend their implementation without careful comparison of their costs and benefits with those of other research projects.

Market-Driven Management

The U.S. Army faces a significant challenge today—how to create a more responsive R&D process that can leverage the relevant network science initiatives of universities and private companies and use them to design and deliver improved network-centric warfighting capabilities to our forces.

As DOD prosecutes the global war on terrorism, it is terrorist (social) networks that are enabling insurgents to innovate, react, and adapt very quickly—in many cases more quickly than U.S. forces. Given the complexity of NCO, the current R&D model is hard-pressed to react, and it is not delivering new insights and capabilities in a sufficiently timely manner to meet Army operational requirements. One only needs to look at the current challenges facing major programs like the Future Combat System (FCS) to understand that the current R&D process needs drastic change (Weiner, 2005).

To field capabilities for NCO more effectively at the strategic, operational, and tactical levels, the Army must develop a model for next-generation R&D, and the evolution of network science may provide a unique opportunity for the Army to transform the R&D process. To make this new model a reality, the Army must think outside the box, leverage existing technology, and attract outstanding researchers both within the Army and in industry and academia.

By combining new thinking, new incentives, and a close partnership among the military, industry, and academia, the new R&D model could leverage existing networks and collaboration technologies to accelerate the development of network science and simultaneously to enhance the R&D process. The new model must take advantage of the rapid proliferation of technology coming from the commercial sector as well as the increasing amount of network science R&D flowing from university research. At the core of the new model will be the innovative use of robust networks and proven, Internet-based business concepts that drive change and increase the speed with which knowledge, research, and technology are delivered.

Companies like Intel, GE, and Oracle, among others, have established network-centric business practices to deliver new technology, products, and services at a rate that keeps them in the forefront of innovation and competition. They do this by keeping their feelers out in the academic, scientific, and technology communities, monitoring evolving technologies, and placing bets on which ones will mature soon and have the most positive market effect. Simultaneously they develop and market products, build market infrastructure to support them, please shareholders, and provide new working capital to continue the process. The Army must adapt and incorporate these industry best practices into its new model, perhaps by creating a networked R&D framework that would engage both industry and academic researchers and more closely integrate them with the Army's in-house R&D resources and stakeholders. The framework could be an eBay-like network model that aggregates ideas and supports a community that rigorously (but rapidly) evaluates them (Surowiecki, 2004).

At the heart of this new model for advancing NCO is the establishment of a national network of virtual centers of excellence at multiple universities, where small groups of researchers act autonomously but are connected to the Army through an R&D coordinating council that also facilitates linkages with private companies. To establish the centers, the Army could implement an initial application process that provides core funding for the researchers. A portion of the funding could be maintained in a focused project fund, which could then be dispensed to research centers that collaborate to address specific Army challenges.

The challenges could be prioritized through an internal application process. Then the Army could post specific problems of interest to which the centers could respond with a brief proposal, estimating time and additional resources needed to address them. The Army would select the most appropriate center to address the challenge and allocate additional funding if necessary. The additional funds might also cover such things as extended data collection, new apparatus needed to perform the task, or programming to generate an interface for a project or even to take an idea to a prototype.

In most cases, members selected to participate in the community would receive base funding to cover their core personnel and other fixed costs. The network of the community would be used to assemble and fund a variety of flexible project teams. In addition to an Army coordinating council, other funding entities (such as other government agencies or private companies) might be invited to participate in this network as sponsors of projects. This mechanism could significantly shorten the normal process between posing a problem or challenge and getting a solution.

Projects could arise in two primary ways. First, the Army coordinating council or another sponsor could post a Request for Proposal (RFP), and various participants from the centers of excellence would submit research proposals. These proposals would be graded by the community of centers, industry participants, and the Army. For large projects, potential participants might also post their own smaller RFPs to assemble teams for subprojects. In either case, project teams could include participants from multiple institutions in the network. When a sponsor wanted help evaluating proposals, it could ask other participants in the network to grade different proposals. Network participants, like those on eBay, would all have a continually updated reputation rating. This buyer-initiated process is similar to that used on sites like Elance, Inc.

A second way for projects to form would be for researchers to post proposals for projects they want to do (or project-related services they are willing to do). In cases of highly desirable projects or services, different sponsors (or subproject leaders) might compete for the services of a person or team. This seller-initiated process is similar to the process used on sites like eBay.

In either case, the centers and researchers who bring worthwhile knowledge, research, and technology to market are funded, and those who don't are not. And the lead time of a year or more that is typical for a grant application process could be reduced to weeks or days.

An example that provides valuable insight is the Networked Centers of Excellence (NCE) program in Canada.[4] NCE mobilizes research talent across Canada and applies it to critical challenges that benefit all Canadians. The program fosters partnerships of business, education, and government to accelerate the exploitation of knowledge, research, and technology and to speed their transfer to products and services that succeed in the marketplace.

One way the Army could organize and apply this new model would be to staff its coordinating council with qualified R&D people (senior officers at the O-5 to O-7 level, including some with warfighting experience) to operate and manage Army and joint R&D projects, as described above. A one-star general could be placed in charge and told that if successful, he/she will be promoted in 3 years. The council could be limited in number to 15-20, mainly to manage the network community described above. It would consist of experienced R&D officers to guide this process, including some O-6 warfighters who would be able to orchestrate or manage the input from the networked community.

The council members would be like Linus Torvald's small group of Linux code approvers or eBay's inner circle of strategic community managers. The Army must give serious thought to how to attract the best minds. Additionally, the Army may want to consider adding to the council one person from industry and one from academia, both of whom are familiar with network research, with what the different centers of excellence can do, and with the cultures of the different communities of researchers.

The council would be given priorities based on the Army's R&D investment strategy, most likely developed by the Deputy Assistant Secretary of the Army for Research and Technology. It would be given authority to cut through red tape and manage both the process and the networked community of contributors.

---

[4]For further information, see NCEnet spring 2004 issue, available at http://www.nce.gc.ca/pubs/ncenet-telerce/spr2004/newsletterspr04_e.htm. Accessed August 19, 2005.

Furthermore, the Army should develop mechanisms for encouraging and recognizing the best and most relevant research. This would involve attracting talented researchers and providing for a constant stream of challenging projects. For continuity, the military officers on the council would be designated "warfighting R&D professionals" and set on an appropriate career path that allows them to develop the next-generation concept of virtual centers of excellence that can meet the Army's critical needs in network science in a short but realistic time frame.

The next-generation model for network science R&D is depicted in Figure E-1. This figure shows the conceptual relationships between the Army coordinating council, university centers of excellence, and industry consortia as described above.

In summary, the next-generation R&D model outlined in this scenario is a new and different approach based on the proven experience of networked organizations like eBay, Intel, and GE. It is based on principles that have worked for many successful companies that needed to get quality products and capabilities to market quickly: Think Big, Start Small, Scale Fast, and Deliver Value. It is this type of next-generation model that can deliver the knowledge, research, and technology that will enable the nation's warfighters to win its wars.

For this approach to work requires strong commitment from the Army and DOD senior leadership and their partners in industry and academia. A small team of the best and brightest Army warfighting R&D specialists committed for 3–4 years and working closely with industry and academia would not only contribute to accelerating the development of the field of network science but also contribute to the prompt and cost-effective delivery of effective NCO capability (DOD, 1999). The Army has an opportunity to assume leadership in developing and implementing this new model. In DOD, NCO must be a joint effort, and so should the new R&D model—after it has been proved in the Army. By moving aggressively to implement this model, the Army can establish itself as the lead in DOD and seize the opportunity to make a significant contribution toward improvement of joint network-centric warfighting capabilities.

The statement of task requests the committee to "identify specific research issues and theoretical, experimental, and practical challenges to advance the field of network science." Briefly stated, one such issue (and major challenge as well) will be to obtain value from the investments that the Army does make to advance network science. In the case of basic research (6.1) alone, relevant challenges are identified in Scenario 1. In the case of the combined basic and applied research (6.1–6.3) projects envisaged in Scenario 2, the challenges depend sensitively on the topics of the research.

To illustrate the scope and scale of next-generation R&D projects with market-driven management, three projects involving the sociological, engineering, and biological areas of network applications were developed by members of the committee as sample projects and are described in following sections.

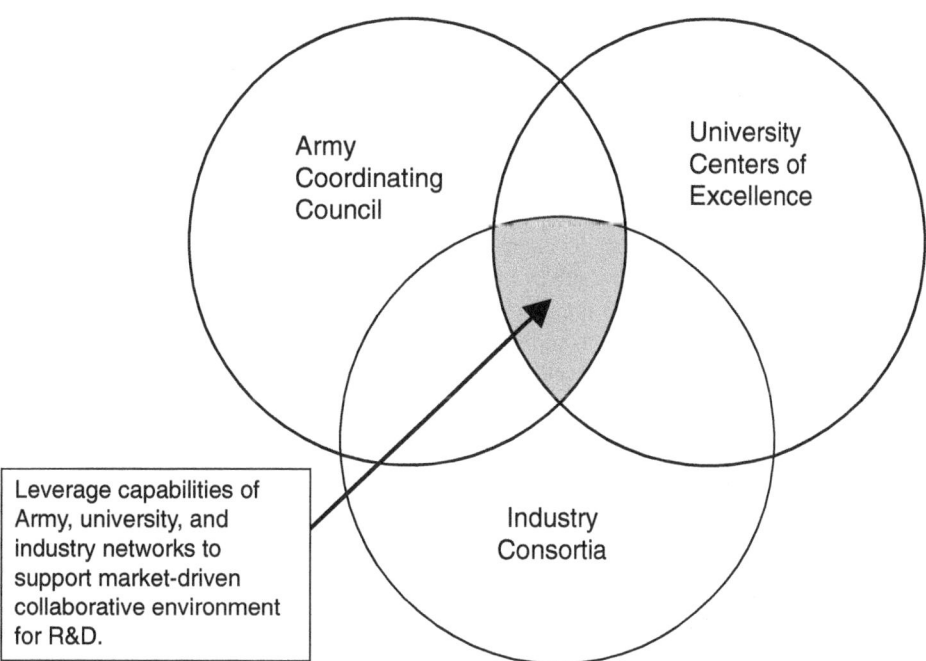

FIGURE E-1 Schematic depiction of next-generation model for Army R&D showing the relationship between the main entities in this model.

Sociological Research: Local Decision Making in a Networked Combat Environment

When network-centric operations are conducted in the field, it is envisioned that squads, platoons, and companies will have access to communications capabilities that provide them with common situational understanding, a common operating picture, information necessary for rapid decision making, and ready communications directly with their commanders and peers. Operations like ground targeting for precision aerial bombardment require that specific items of information be made accessible to specific elements on (or above) the battle. Experiments have been conducted to determine whether access to this information actually increases combat effectiveness, and if it does, how. This proposal is for an exploration of decision making in an information-rich combat environment via the design and conduct of field tests to assess how Army personnel might adapt their processes for searching for information and for decision making to make optimal use of newly acquired information resources.

This problem is of profound interest to the Army because an understanding of the impact of incremental increases in information resources on the search and decision making processes of field units is central to the utilization of these resources to improve the combat effectiveness of these units. This is especially true if the units are expected to fight under nontraditional circumstances like urban guerilla combat.

The proposed approach to this problem is to extend and to adapt current models of networked group decision making to examine how the availability of NCO capabilities might modify the decision-making processes used by combat units in the field and hence the effectiveness of these units (Watts, 2003). The current models apply more to simpler, more generic situations than to the complex situation of decision making in combat. Thus, three lines of investigation would be initiated. First, extensions of the models would be examined to assess the extent to which they can be applied to social situations characterized by information exchange between remotely located small groups collaborating to reach a joint decision. Second, examples of specific unit decisions in field situations would be reviewed to assess if a subset of the decisions could be isolated and subjected to the discipline of constructing validated models. Third, the availability and accessibility of NCO information in field settings would be explored to assess the types and amount of information that realistically could be expected under combat situations. Following these initial lines of investigation, an assessment would be made of the feasibility of combining the results to design and field test a revised set of decisioning processes that showed promise of using the available NCO information assets to improve combat effectiveness. At this stage a decision would be made on whether the prospect of useful results was good enough that the project could usefully be scaled up to support an actionable technology investment option to develop elements of training modules or specifications for the user interface to and functionality of NCO capabilities.

The committee envisages that the models would be constructed and exercised by leading university groups. Their test and validation by means of field experiments and simulations would be performed at Army facilities using primarily Army personnel. These two groups must work together intimately, because the current state of the art in modeling obviously does not extend to the complexities of realistic combat decisioning. Therefore, success of this project depends on an exceptional level of creativity both in the extension of current models and in the identification of specific field-relevant situations in which they could reasonably be tested.

The responsibilities of the university personnel would be to explore the range of behaviors described by current models and their extensions and to work with Army personnel to assess their applicability to unit decision making in an NCO environment. The responsibilities of the Army personnel include management of the program, collaboration with their university partners to assess the applicability of the models to NCO, design and conduct of tests, participation in the analysis and interpretation of test results, and preparation of recommendations to the Army for changes in unit organization and training processes based on the results.

The Army could make this program attractive to researchers at leading universities by making 5-year grants for the work, facilitating the participation of non-U.S. citizens, allowing for timely publication of the results of the 6.1 portions of the program, providing access to state-of-the-art computing and Web collaboration facilities, and making the competition for the grants sufficiently rigorous that only top-notch university personnel could participate.

Special capabilities that would be required include enough computer power to exercise the models adequately and simulation and test facilities for testing models of NCO information availability and usage.

This proposal envisages an intimate collaboration between Army and university personnel in all aspects of planning and executing the program. The project would start with a workshop in which the goals and approach of the project are defined and procedures for collaboration clarified. The principal university personnel involved would be expected to devote at least 3 months each year to the project, with up to a full year temporary assignment possible during the conduct of the experiments. Army personnel would be made available to give participating university groups extended briefings on the realities of decision making under combat conditions. Efforts would be made to keep the assignment of the leaders (both Army and university) of the program stable so that good personal and professional relationships would be established.

This project is expected to differ from traditional Army 6.1 programs in focus, intensity, and style. It would construct, validate, and refine models of combat decision making in an NCO environment that would be usable by the Army in determining the kind of information that is needed and in training Soldiers. The research is expected to be performed in the manner of a team social network, and the project would be intense enough to require full-time involvement of participants for extended periods. The university and Army partners will construct a validated model to assess if specific items of information envisaged to be available in an NCO environment can be used to create a decision-making process that demonstrably improves combat effectiveness.

The proposed project also differs from projects sponsored by funding agencies like NSF or the Department of Energy by virtue of being a basic research project in model building and validation that is embedded in a design, test, and evaluation environment that would be suitable for the observation and description of human behavior in particular network-centric operations performed by the Army.

Engineering Research: Large-Area Agent Monitoring Network

The goal of this project would be to develop and demonstrate the ability to create algorithms for communications, sensor fusion, and motion control that can maintain an accurate representation of the location, role, and intent of a distributed set of 10,000+ agents in a distributed geographic area. The sensors would range from high-resolution fixed assets (e.g., satellites), to sensors on platforms whose motion can be dedicated to data collection (e.g., unmanned aerial vehicles [UAVs]), to sensors whose motion is under limited control (e.g., sensors attached to already moving objects or fixed in place). Connectivity between the sensors varies based on local conditions: Some sensors are sporadically available, others are mostly available but moving and changing the network topology, and still others are permanently attached to the network.

A civilian example of this capability would be to maintain a dynamic population density estimate across a city the size of Washington, D.C., that is accurate to within +/− 100 people per square kilometer and to use this information to control the flow of individuals to safety shelters inside or outside the city through a heterogeneous set of transportation systems (cars, trains, buses, walking, etc.). Sensors would include fixed sensors deployed on buildings and in public areas, mobile sensors deployed on vehicles (some of which are dedicated to collecting information, others of which have other primary tasks), and personal sensors carried by emergency personnel.

A military example of this capability would be to maintain an accurate representation of an urban area that included information on the location of (possibly moving) military assets, on the likely location of specific individuals and groups of military interest, on the status of civilian and military infrastructure (roads, communications, power, etc.), and other information relevant to a military operation. Information sources would include dedicated satellite and UAV data, sensors attached to military vehicles operating in the area, and fixed and mobile sensors (including cameras, vibration sensors, communications taps, and power monitoring equipment) placed by military personnel.

Current solutions to this problem involve either trying to centralize all information (as, for example, in an air operations center) or maintaining sparsely interconnected information sources (different military units maintain their own operating pictures, and these are synchronized at a high level and with limited fidelity). The main obstacle to better integration of information is not the communications bandwidth available, but rather our ability to design, build, and operate complex, networked systems at this scale. Traditional approaches to analyzing the dynamics and function of such networks are not likely to provide the insights necessary to design robust, scalable solutions to this problem.

One approach to solving this problem while advancing the field of network science would be to develop a sequence of annual experiments in which new methods and approaches could be tested in a highly instrumented environment, roughly like the military exercises run at the National Training Center (NTC). The Army would define and pay for running the experiments and would provide access to the data collected on various nodes. Universities and companies would demonstrate technologies at the experiments, including forming coalitions to integrate their technologies into a complete solution. A resident military group—patterned after, say, the opposing force (OPFOR) at the NTC—would role play the opposition and try to exploit the technology so that it provided incorrect information about what is going on. All data from the experiments would be made available to the research community at some point after completion of the experiments.

To implement this set of experiments, the Army would have to first identify (or construct) a site and to install instrumentation to allow monitoring of the operation. The type of facility envisaged by the committee would be similar to the Army's NTC at Fort Irwin, which already has many of the types of instrumentation that would be required to keep track of the movement of and communications between a large number of troops operating in a military exercise.

Finally, the Army would have to develop (or contract someone to develop) a software and hardware framework that defines the interfaces to the portions of the experiment that are fixed (e.g., the vehicles, sensors, and human-interface hardware). The interface specifications and a simulation and/or test suite would be made available to all prospective participants in the exercise. These specifications should be focused on exploring the role of network effects in operations and need to represent current military capabilities, systems, or platforms.

A possible model for this framework is the Open Control Platform (OCP), developed by Boeing as part of the Software-Enabled Control (SEC) program of the Defense

Advanced Research Projects Agency (DARPA). The OCP allowed researchers from academia and industry to run a set of experiments on a test bed consisting of an F-16 and a T-33. Each research site developed its own code, then tested and verified it in simulation. Then the code was run on the experiment, with full data provided to the users at the end of the runs. Operation of the entire SEC final experiment was the responsibility of Boeing. Although this program was much smaller in overall scope than what the committee is proposing here, it provides one example of how to allow many users to access a common test facility.

The experiments must be sufficiently large and complex that a systems-level approach is required to succeed. The committee anticipates that an experiment would involve between 10 and 100 vehicles, hundreds of troops (each connected to the network), and 10,000+ individual sensors and information sources. The venue of the experiment would have a like number of individual agents that needed to be tracked and characterized, as described in the civilian and military applications above.

The participants would be teams of researchers, technologists, and engineers from academia and industry. University researchers would provide new ideas for methods to integrate information, implemented as protocols, algorithms, and theory. They would develop software for communication protocols, sensor fusion, control, etc. Industry would be responsible for overall systems integration as well as implementation of advanced methods for information aggregation in a networked environment. The Army would manage the operation but also insert new technologies from Small Business Innovation Research (SBIR) and other entities receiving 6.2 funding under their control. Several recent examples of such programs, such as the DARPA SEC and Bio-Spice programs, could be used as models for how the teams would work together.

The program would be structured to attract participants. The availability of large-scale experimental infrastructure and data sets for further research would motivate participation by universities, since this capability would be well beyond what is typically available in an academic environment. It would also be important to provide a mechanism for long-term support of fundamental research in network science, perhaps by engaging a dedicated set of researchers to study the overall experiment and provide a running assessment of the adequacy of the science to explain the results of the exercise. In addition to allowing exploration of engineering aspects of network science, this type of environment could also be used to explore the social aspects of networks in military operations—for example, the impact of changes in the network on the tactics used by friendly and enemy units.

Biological Research: Field Biological Threat Assessment

Military personnel are deployed widely in different areas of the globe, are exposed to the different environments and diseases at these locations, and are then, often quickly, redeployed elsewhere in the world. In addition, there is a threat of directed biological attack on the force during combat. These intense conditions can lead to the rapid infection of troops and the spread of agents across the world if not immediately detected and neutralized. Such a threat is also encountered in civilian situations, where economic needs might drive the invasion of remote locations, such as biologically diverse rain forests to collect medicines or wilderness area to drill for oil. Other such situations might entail the exchange of bacterially and virally dense sea waters through ballast uptake and dumping by commercial or tourist ships. Each of these cases involves the introduction of infective biological systems from once-isolated remote locations to places where individuals have not built immunity to them.

The mobility of the military, or even civilian, participants means infections could be brought from the site of interaction to the wider community. Avian influenza provides a prime illustration of the danger posed (see Box E-1) and underscores the importance of having in place a well-informed surveillance and decision structure to detect and respond to emerging biological threats.

The goals of this project would be to develop a focused capability to monitor the biological load of troops prone to rapid redeployment, to demonstrate a capability to detect and identify the source of new biological threats to soldier health, and to provide recommendations for the control of such threats to individuals and populations. A longer term goal would be to turn this information into sufficient knowledge about the biological organism at the molecular level to allow rapid development and deployment of pharmaceutical solutions. For the purposes of this report the committee uses as examples existing viral threats such as human immunodeficiency virus (HIV-1), influenza, and West Nile virus and the emerging threat of avian flu. Similar considerations apply, however, to bacterial threats such as anthrax, tularensis, and leishmania (a particular problem in the Middle Eastern theater).

This overall assessment effort would require a network of monitoring stations reporting on samples from the environment and patients, looking for prevalence of the threat agent and patient response to the agent in terms of morbidity and virulence indices. Agents would be rapidly sequenced and typed. These sequences would be related to their virulences and used to track the spread of a particular strain from location to location. Models of the viral infection and transmission process based on the sequence and prior measurements of similar strains would inform network models of the spread and mutation of these viruses. These models, in turn, would enable rapid risk assessments when an infection appears at a given site.

This biology project links with the social and engineering network sample projects presented above. Data from arrays of sensors in the environment and on individual soldiers would need to be fused into a database of biological and

> **BOX E-1**
> **Case Study from the World Health Organization: Avian Influenza**
>
> Avian influenza is an infectious disease of birds caused by the Type A strain of influenza. Nearly all birds are susceptible to it, and different strains have widely differing virulences, with some strains being highly contagious and able to cause severe symptoms, death, or dangerous epidemics. There are at least 15 subtypes of avian influenza that provide an extensive pool of virus. Contact between domestic and migratory birds transmits and mixes these pools, and research has shown that even viruses of low pathogenicity can, after circulating for only a short time, mutate into highly pathogenic viruses. Early detection of these virulent strains and strict quarantine are the most effective means of control. Good sanitation helps in prevention, but virus can exist in the environment for long periods, especially in cold regions. In the absence of prompt control measures backed by excellent surveillance, epidemics can last for years.
>
> One of the most worrisome aspects of these viruses is their ability to efficiently recombine their genetic material with that of other viruses to cause an antigenic shift, creating variants less likely to be halted by the immune systems of the birds. These viruses rarely hop species (except to pigs), but there have been a number of recent cases of the avian flu hopping from birds to humans. These cases are so disturbing that a 1997 outbreak of one highly pathogenic strain (H5N1) in Hong Kong poultry that directly spread to 18 humans, 6 of whom died, caused the government to destroy within 3 days the entire poultry population of 1.5 million birds. This action likely averted a large-scale spread to humans and a pandemic. The infected persons were often otherwise healthy, and there seemed to be no age preference. Since then there have been a number of other outbreaks in southern China, Viet Nam, and Korea.
>
> The H5N1 strain is of particular concern because it has a propensity to acquire genes from viruses that infect other animal species, increasing the probability that these new strains might become virulent to humans. Testing and diagnosis for the virus is rapid and reliable, and treatment is the same as for human flu. Vaccines can be effective, although only temporarily. Antivirals often work, but they are expensive and in short supply. The upshot is that rapid detection, estimation of the threat, and good command and control, such as was demonstrated in the Hong Kong poultry massacre, are important means for containing this possible source of pandemic.
>
> SOURCE: Adapted from a document found at http://www.who.int/csr/don/2004_01_15/en/.

geographical information. This database would be provided with analytical and decision management tools that assess the risk associated with detection of a new strain of virus or infection event. The information would need to be coordinated from forward field laboratories to commanders responsible for combat operations, especially troop deployment.

A pilot study in this area would be designed to cooperate with Marine and U.S. Navy SeaBee medical aid stations (MASs) already engaged in disease and noncombat illness surveillance, as well as with the global disease tracking efforts of the Centers for Disease Control (CDC) and the World Health Organization (WHO). The MASs could be charged with monitoring soldiers for a panel of viral species and would be outfitted to rapidly sequence isolates. Perhaps a field team would be dispensed to sequence similar viruses in local fauna and environments. Sequences would be filed in databases along with soldier data (health and some form of identifier for tracking where individuals are stationed and how they are moved from place to place). These databases would be integrated with, for example, the influenza database at Los Alamos National Laboratory.[5] A university team would work on relating sequence, virulence, and geographic information into a model of the sequence determinants of different virulence and epidemiological parameters, such as transmissibility and the rate of mutation. Working with a DOD team, the university team would develop structured epidemiological models of the spread and of new viral infections in the troop population based on models of troop movement and the viral dynamics. An Army team would then focus on modeling decision processes under different scenarios in order to optimize containment of any viral threat.

This project involves four major challenges: (1) choosing the scale, breadth, and accuracy of the data needed for construction and parameterization of the network models, (2) developing a modeling framework relating data at the viral sequence level to the level of whole populations of infected individuals, (3) creating modeling tools that take into account the unique ability of live biological threat agents to adapt to and mutate around human interventions, and (4) developing a decision support model that prescribes an effective course of medical and social intervention to quell a possible viral outbreak. This work faces the challenges generally faced by multimodal, multiscale network analysis—how to place the sensors (in this case medical and environmental surveillance); how to fuse their information into network models at both the epidemiological and molecular

---

[5] For further information, see http://www.flu.lanl.gov. Accessed August 19, 2005.

levels; and how to use these network models to direct distribution of resources and to design interventions—and requires network science at all scales. Since coordination and communication among the teams is paramount to the success of a project at this scale, the online infrastructure and computational resources would be an early focus.

An initial coordination team would identify the key existing infrastructure for achieving the project goals and would establish contact with the requisite field sites and field military personnel. The team would develop a knowledge of the key data trackers in the WHO and CDC. It would identify candidates to participate in the program. Key government players and invited participants would attend a workshop to discuss what technology, resources, and organization would be necessary for success of the program. A subset of this group would form a project team and create the final 5-year project design. Dedicated team leaders would be chosen from the Army, other government agencies, and nonmilitary research groups.

The scale of the work, the diversity of participating agencies, and the use of military field data make this a project outside the scope of most funding agencies. The committee expects that the basic science to be accomplished as well as the important societal and military applications would make this project attractive to the scientific community at large.

SCENARIO 3, CREATING A ROBUST NCW/NCO CAPABILITY

The statement of task instructs the committee to "recommend those relevant research areas that the Army should invest in to enable progress toward achieving network-centric warfare capabilities." When the committee examined the literature on this topic from the DOD OFT, it discovered that the concept of "network-centric warfare" had been superseded by the concept of "network-centric operations," as described in a conceptual framework document published on the OFT Web site[6] (Cebrowski and Garstka, 1998; Garstka and Alberts, 2004). In interviews and discussions with representatives of the Army and DOD, committee members learned that opinions on NCW and NCO varied widely and were substantial, not just a matter of nomenclature. Moreover, the literature on the topic is dynamic, with new reports and publications frequent. Since this report is intended as an archival document, the committee elected to use the published conceptual framework description version 2.0 (Garstka and Alberts, 2004) as its point of reference.

In Scenario 3, the committee adopts a national point of view. Its purpose is to ask what the nation must do if the strategic vision of NCO is to become a reality. Investment in research by the Army is a part of Scenario 3, albeit a modest one.

Transforming the U.S. military from its current state to that envisaged for NCO, as described by Garstka and Alberts (2004), is probably the most complex undertaking in the history of the U.S. government. Its achievement would arguably be comparable to the successful pursuit of World War II or the cold war with the Soviet Union. It is a long-term, difficult, costly, and risky undertaking. It is not clear to the committee that the difficulty, cost, and risk associated with this notion have been communicated effectively to the senior management of the Army and DOD. In Scenario 3, the committee emphasizes the magnitude of the undertaking and sets forth the view that significant new activities will be required to accomplish it.

Think first of designing the most complex weapons system yet: say, a large aircraft carrier. Add to this the complications in the physical domain associated with, for example, secure, reliable wireless communications via satellite to soldiers on a mobile battlefield. In the information domain, add the hardware and software challenges associated with storage, search, and retrieval of orders of magnitude more data than have ever before been processed in real time, as well as the challenges associated with ensuring the security and reliability of these data. In the cognitive domain, add issues associated with processing information from all three services by a junior officer at a local (mobile) workstation. In the social domain, add the complications of orchestrating the decision-making process in this information-rich, real-time environment and the issues associated with tactics and training to use all this information-processing capability. In all seriousness, the challenge seems more like science fiction than like science and technology that can be delivered up by the R&D operations of the military services as currently constituted or even by commercial R&D.

The committee regards it as highly unlikely that existing methods of designing and procuring weapons systems will be adequate to accomplish this monumental undertaking. Current experience in the services themselves supports this point of view (Brewin, 2005). The committee regards the task of converting the current state of the U.S. military to the vision articulated for NCO as vastly more challenging than seems to be appreciated.

Not only is the task dauntingly complex, the knowledge necessary to accomplish it does not even exist. In similar cases—the Manhattan Project and the initial days of NASA come to mind—a focused, long-term national initiative was required, and it seems likely that something similar will be required in this case also. Thus, in Scenario 3 the United States undertakes a focused national initiative, comparable in scope to the Manhattan Project, to design and deploy NCO capabilities as described in the conceptual framework document version 2.0 for all the military services during the coming decade.

---

[6]For further information, see http://www.oft.osd.mil/. Accessed August 19, 2005.

Insights from Network Thinking

What insight does network thinking offer on this subject? First, a network organization (as opposed to a hierarchical one) is required because the organization must learn how to do the job on the job.[7] A critical function is specifying the architecture of the interacting networks in the various domains, setting the interfaces between them, and monitoring successive waves of implementation to ensure consistency and learning. This could be accomplished by a senior executive office that also managed the program itself. Underneath this top layer of the network is a project management layer responsible for converting the architecture into a series of projects that implement and test various elements of the architecture. Think of this layer as the middleware of the organization. Beneath this is a network layer of vendors and military logistic organizations that provide and source the material for operational commands. A model for the vendor organization might be Sematech. A model for the logistics organization might be the nuclear Navy. The important points are that a new network organizational structure and work process is required to create a capability of this complexity and that the scale of the activity is larger than the Army alone; it must be national, or at least DOD-wide.

Another insight is that the design and implementation of NCO capabilities are on the same scale and of the same complexity as their use. The same principles of better, faster decisions by means of information sharing apply. Thus, the environment in which the capabilities are designed can be regarded as a test bed for many of the expected capabilities themselves in the information, cognitive, and social domains. Taking advantage of this insight could make the overall job less onerous and should be an explicit consideration in initiatives to create NCO capabilities.

Synopsis

The committee was not tasked to resolve the issues raised in Scenario 3 but considers their resolution to be of paramount national urgency. The committee stresses that the knowledge of networks that we possess today is not adequate to design predictable, secure, robust global networks. Members heard presentations and read reports on how the "transformation" to a future force capable of NCO is not likely to be achieved by traditional approaches to creating technology. They came to recognize that the policies and practices currently used to procure these capabilities do not take into consideration the uncertainties inherent in the current state of understanding the design and implementation of complex networks. The purpose of Scenario 3, then, is to emphasize that the task of designing, testing, and operating the envisaged NCO capabilities is of an exceedingly high order of complexity and should be approached as seriously as the Manhattan Project or NASA's race to the moon.

The committee would be remiss in its responsibilities if it failed to note the essential urgency and profound difficulty of this task. The chances of delivering NCO capabilities in a timely and affordable way would be greatly increased by a focused national initiative combining the initiatives of all services under central leadership, to respond successfully to the diverse challenges of future warfare.

REFERENCES

Albert, R., and A.L. Barabási. 2002. Statistical mechanics of complex networks. Review of Modern Physics 74(1): 47–97.

Albert, R., H. Jeong, and A.L. Barabási. 2000. Error and attack tolerance of complex networks. Nature 406(6794): 378–382.

Almaas, E., B. Kovacs, T. Vicsek, Z.N. Oltvai, and A.L. Barabási. 2004. Global organization of metabolic fluxes in the bacterium Escherichia coli. Nature 427(6977): 839–843.

Amaral, L.A.N., A. Barrat, A.L. Barabási, G. Caldarelli, P. De Los Rios, A. Erzan, B. Kahng, R. Mantegna, J.F.F. Mendes, R. Pastor-Satorras, and A. Vespignani. 2004. Virtual roundtable on ten leading questions for network research. European Physical Journal B 38(2): 143–145.

Anderson, R.M., and R.M. May. 1992. Infectious Disease of Humans: Dynamics and Control. Oxford, England: Oxford University Press/Oxford Science Publications.

Arquilla, J., and D. Ronfeldt. 2001. Networks and Netwars: The Future of Terror, Crime, and Militancy. Santa Monica, Calif.: RAND.

Barabási, A.L. 2003. Linked. New York, N.Y.: Plume, a member of the Penguin Group, Inc.

Barabási, A.L., and Z.N. Oltvai. 2004. Network biology: Understanding the cell's functional organization. Nature Reviews: Genetics 5(2): 101–114.

Barrat, A., M. Barthélemy, R. Pastor-Satorras, and A. Vespignani. 2004. The architecture of complex weighted networks. Proceedings of the National Academy of Sciences of the United States of America 101(11): 3747–3752.

Ben-Naim, E., H. Frauenfelder, and Z. Toroczkai. 2004. Complex Networks. New York, N.Y.: Springer-Verlag.

Bornholdt, S., and H.G. Schuster, eds. 2003. Handbook of Graphs and Networks: From the Genome to the Internet. Weinheim, Berlin: Wiley-VCH.

Brewin, R. 2005. DOD Mulls Network Coordination. Available at http://www.fcw.com/article88939-05-23-05-Print/. Accessed May 31, 2005.

Cebrowski, A., and J. Garstka. 1998. Network centric warfare. Proceedings of the United States Naval Institute 24: 28–35.

Cohen, R., K. Reez, D. Ben-Avrahan, and S. Havlin. 2000. Resilience of the Internet to random breakdowns. Physical Review Letters 85(21): 4626–4628.

Cohen, R., K. Reez, D. Ben-Avrahan, and S. Havlin. 2001. Breakdown of the Internet under intentional attack. Physical Review Letters 86(16): 3682–3685.

Department of Defense (DOD). 1999. DOD Financial Management Regulation Volume H.

Dodds, P.S., and D.J. Watts. 2004. Universal behavior in a generalized model of contagion. Physical Review Letters 92(21): 218701.

Dodds, P.S., and D.J. Watts. 2005. A generalized model of social and biological contagion. Journal of Theoretical Biology 232(4): 587–604.

Dodds, P.S., D.J. Watts, and C.F. Sabel. 2003. Information exchange and the robustness of organizational networks. Proceedings of the National Academy of Sciences of the United States of America 100(21): 12516–12521.

---

[7]C.F. Sabel, professor of law and social science, Columbia Law School, "Theory of a real-time revolution," briefing to the 19th EGOS Colloquium, Copenhagen, July 2003.

Dorogovtsev, S.N., and J.F.F. Mendes. 2003. Evolution of Networks: From Biological Nets to the Internet and WWW. Oxford, England: Oxford University Press.

Garstka, J., and D. Alberts. 2004. Network Centric Operations Conceptual Framework Version 2.0. Vienna, Va.: Evidence Based Research, Inc.

Gladwell, M. 2000. The Tipping Point: How Little Things Can Make a Big Difference. New York, N.Y.: Little, Brown and Company.

Helper, S., J.P. MacDuffie, and C. Sabel. 2000. Pragmatic collaborations: Advancing knowledge while controlling opportunism. Industrial and Corporate Change 9(3): 443–483.

Huckfeldt, R., P.E. Johnson, and A. Sprague. 2004. Political Disagreement: The Survival of Diverse Opinions within Communication Networks. Cambridge, England: Cambridge University Press.

Katz, E., and P.F. Lazarsfeld. 1955. Personal Influence: The Part Played by People in the Flow of Mass Communications. Glencoe, Ill.: Free Press.

Newman, M.E.J. 2003. The structure and function of complex networks. SIAM Review 45(2): 167–256.

Newman, M.E.J., A.L. Barabási, and D. Watts, eds. 2005. The Structure and Dynamics of Networks. Princeton, N.J.: Princeton University Press.

Pastor-Satorras, R., and A. Vespignani. 2004. Evolution and Structure of the Internet: A Statistical Physics Approach. Cambridge, England: Cambridge University Press.

Rogers, E.M. 1995. Diffusion of innovations. New York, N.Y.: Free Press.

Surowiecki, J. 2004. The Wisdom of Crowds: Why the Many Are Smarter Than the Few and How Collective Wisdom Shapes Business, Economies, Societies and Nations. New York, N.Y.: Doubleday.

Tanenbaum, A.S. 2003. Computer Networks, 4th edition. Upper Saddle River, N.J.: Prentice Hall PTR.

Watts, D.J. 2002. A simple model of information cascades on random networks. Proceedings of the National Academy of Sciences of the United States of America 99(9): 5766–5771.

Watts, D.J. 2003. Six Degrees: The Science of a Connected Age. New York, N.Y.: Norton.

Watts, D.J. 2004. The "new" science of networks. Annual Review of Sociology 30(1): 243–270.

Weiner, T. 2005. "Drive to build high-tech Army hits cost snags." *New York Times*, March 28, 2005.

F

Recommended Reading List

Ahmadjian, C.L., and J.R. Lincoln. 2001. Keiretsu, governance, and learning: Case studies in change from the Japanese automotive industry. Organization Science 12(6): 683–701.

Alberts, D.S., and R.E. Hayes. 2003. Power to the Edge: Command and Control in the Information Age. Washington, D.C.: CCRP Publication Series.

Alberts, D.S., and T.J. Czerwinski, eds. 1997. Complexity, Global Politics and National Security. Washington, D.C.: National Defense University Press.

Amin, M. 2002. Modeling and control of complex interactive networks. IEEE Control Systems Magazine: 22–27.

Arquilla, J., and D. Ronfeldt. 2000. Swarming and the Future of Combat. Santa Monica, Calif.: RAND National Defense Research Institute.

Artzy-Randrup, Y., S.J. Fleishman, N. Ben-Tal, and L. Stone. 2004. Comment on "Network motifs: Simple building blocks of complex networks" and "Superfamilies of evolved and designed networks." Science 305(5687): 1107c.

Barabási, A.L. Linked. 2003. New York, N.Y.: Plume, a member of the Penguin Group, Inc.

Barabási, A.L., and Z.N. Oltvai. 2004. Network biology: Understanding the cell's functional organization. Nature Reviews: Genetics 5(2): 101–114.

Berkowitz, B. 2003. The New Face of War: How War Will Be Fought in the 21st Century. New York, N.Y.: Simon and Schuster.

Bower, J.M., and H. Bolouri, eds. 2001. Computational Modeling of Generic and Biochemical Networks. Cambridge, Mass.: MIT Press.

Buchanan, M. 2002. Nexus: Small Worlds and the Groundbreaking Science of Networks. New York, N.Y.: Norton.

Bushnell, L.G. 2001. Networks and control. IEEE Control Systems Magazine (1): 22–23.

Carley, K.M. 1995. Communication technologies and their effect on cultural homogeneity, consensus, and the diffusion of new ideas. Sociological Perspectives 38(4): 547–571.

Carley, K.M. 2001. Smart Agents and Organizations of the Future. Thousand Oaks, Calif.: Sage Publications.

Carley, K.M. 2003. Dynamic network analysis. Pp. 133–145 in Dynamic Social Network Modeling and Analysis: Workshop Summary and Papers. Washington, D.C.: National Academy Press.

Carley, K.M., J.S. Lee, and D. Krackhardt. 2001. Destabilizing networks. Connections 24(3): 31–34.

Clippinger, J.H. Human Nature and Social Networks. Available at http://www.dodccrp.org. Forthcoming.

Davis, S., and C. Meyer. 1998. Blur: The Speed of Change in the Connected Economy. New York, N.Y.: Warner Books.

Davis, S., and C. Meyer. 2003. It's Alive: The Coming Convergence of Information, Biology, and Business. Chapter 3 and Chapter 7. New York, N.Y.: Crown Business.

Epstein, J.M. 2002. Modeling civil violence: An agent-based computational approach. Proceedings of the National Academy of Sciences of the United States of America (PNAS) 99(Suppl. 3): 7243–7250.

Ferber, D. 2004. Synthetic biology: Microbes made to order. Science 303(5655): 158–161.

Fukuyama, F. 1995. Trust: The Social Virtues and the Creation of Prosperity. New York, N.Y.: Free Press.

Gaddis, J. 2004. The Landscape of History: How Historians Map the Past. Vienna, Va.: Evidence Based Research, Inc.

Garstka, J., and D. Alberts. 2003. Network Centric Operations Conceptual Framework Version 2.0. Vienna, Va.: Evidence Based Research, Inc.

Gerchman, Y., and R. Weiss. 2004. Teaching bacteria a new language. Proceedings of the National Academy of Sciences of the United States of America 101(8): 2221–2222.

Goldsmith, S., and W.D. Eggers. 2004. Governing by Network: CIOs and the New Public Sector. Washington, D.C., and Cambridge, Mass.: Brookings Institution Press and Innovations in American Government Program at the John F. Kennedy School of Government at Harvard University.

Hammes, T.X. 2004. The Sling and the Stone: On War in the 21st Century. Osceola, Wis.: Zenith Press.

Hoffman, F.G., and G. Horne, eds. 1998. Maneuver Warfare Science. Quantico, Va.: U.S. Marine Corps Combat Development Command.

Horn, P. 2005. The new discipline of services science. Business Week. Available at http://ww.businessweek.com/technology/content/jan2005/tc20050121_8020.htm?campaign_id=nws_techn_jan25&link_position=link10. Last accessed on February 28, 2005.

Ilachinksi, A. 2004. Artificial War: Multiagent-Based Simulation of Combat. Hackensack, N.J.: World Scientific Press.

James, G.E. 1995. Chaos Theory: The Essentials for Military Application. Newport, R.I.: Naval War College Press.

Jervis, R. 2001. System Effects: Complexity in Political and Social Life. Princeton, N.J.: Princeton University Press.

Keller, E.F. 2002. Making Sense of Life: Explaining Biological Development with Models, Metaphors and Machines. Cambridge, Mass.: Harvard University Press.

Keller, E.F. Revisiting "scale-free" networks. BioEssays. In press.

Kelly, K. 1999. New Rules for the New Economy. New York, N.Y.: Penguin Group.

Kobayashi, H., M. Kærn, M. Araki, K. Chung, T.S. Gardner, C.R. Cantor, and J.J. Collins. 2004. Programmable cells: Interfacing natural and engineered gene networks. Proceedings of the National Academy of Sciences of the United States of America 101(22): 8414–8419.

Looger, L.L., M.A. Dwyer, J.J. Smith, and H.W. Hellinga. 2003. Computational design of receptor and sensor proteins with novel functions. Nature 423(6936): 185–189.

Lun, L., D. Alderson, W. Willinger, and J. Doyle. 2004. A first-principles approach to understanding the Internet's router-level topology. Proceedings of SIGCOMM 04.

Malone, T.W. 2004. The Future of Work: How the New Order of Business Will Shape Your Organization, Your Management Style and Your Life. Cambridge, Mass.: Harvard Business School Press.

Murray, R.M., ed. 2003. Control in an Information Rich World, Report of the Panel on Future Directions in Control, Dynamics, and Systems. Philadelphia, Pa.: Society for Industrial and Applied Mathematics.

National Research Council (NRC). 2001. Opportunities in Biotechnology for Future Army Applications. Washington, D.C.: National Academy Press.

Newman, M.E.J. 2001. The structure of scientific collaboration networks. Proceedings of the National Academy of Sciences of the United States of America 98(2): 404–409.

Newman, M.E.J. 2003. The structure and function of complex networks. SIAM Review 45(2): 167–256.

Parmentola, J.A. 2004. Army transformation: Paradigm-shifting capabilities through biotechnology. The Bridge: Linking Engineering and Society 34(3): 33–39.

Perez, C. 2002. Technological Revolutions and Financial Capital: The Dynamics of Bubbles and Golden Ages. Cheltenham, England: Edward Elgar Publishers.

Reed, D.P. 1999. Weapon of Math Destruction. Available at http://www.contextmag.com/archives/199903/DigitalStrategy.asp/. Last accessed on March 23, 2005.

Rheingold, H. 2002. Smart Mobs: The Next Social Revolution. Cambridge, Mass.: MA Basic Books.

Sable, C.F. 2003. Theory of a real time revolution. Paper presented at the 19th EGOS Colloquium. July 3–5. Forthcoming.

Scott, J. 2000. Social Network Analysis: A handbook. Thousand Oaks, Calif.: SAGE Publications.

Silver, P., and J. Way. 2004. The potential for synthetic biology. The Scientist 18(18): 30–31.

Smith, M.A. 1998. Communities in Cyberspace. Oxford, England: Routledge.

Strogatz, S.H. 2001. Exploring complex networks. Nature 410(6825): 268–276.

Strogatz, S.H. 2003. SYNC: The Emerging Science of Spontaneous Order. Pp. 229–259. New York, N.Y.: Theia Books.

Tanenbaum, A.S. 2003. Computer Networks, 4th edition. Upper Saddle River, N.J.: Prentice Hall PTR.

U.S. Marine Corps (USMC). 1996. Control and Command. Distribution Number 142 000001 00. Washington, D.C.

USMC. 1997. Warfighting. Distribution Number 142 000006 00. Washington, D.C.

Von Bertalanffy, L. 1976. General System Theory: Foundations, Development, Applications. New York, N.Y.: George Braziller, Inc.

Waldrop, M.M. 1992. Complexity: The Emerging Science at the Edge of Order and Chaos. New York, N.Y.: Simon and Schuster.

Watts, D.J. 2003. Six Degrees: The Science of a Connected Age. New York, N.Y.: W.W. Norton.

Watts, D.J. 2004. The "new" science of networks. Annual Review of Sociology 30(1): 243–270.

Watts, D.J., and S.H. Strogatz. 1998. Collective dynamics of "small-world" networks. Nature 393(6666): 440–442.

Wellman, B. 1996. Computer networks as social networks: Collaborative work, telework, and virtual community. Annual Review of Sociology 22(1): 213–238.

Wellman, B. 2001. Computer networks as social networks. Science 293(5537): 2031–2034.

Wellman, B. 2001. Physical place and cyberplace: The rise of personalized networks. International Journal of Urban and Regional Research 25(2): 227–252.

West, G.B., and J.H. Brown. 2004. Life's universal scaling laws. Physics Today 57(9): 36–43.

Wilson, C. 2004. Network Centric Warfare: Background and Oversight Issues for Congress, Order Code RL32411. Washington, D.C.: Congressional Research Service, Library of Congress.

Wolkenhauer, O., B.K. Ghosh, and K.H Cho. 2004. Control and coordination in biochemical networks. IEEE Control Systems Magazine (4): 30–34.